THE HOME OF THE STARS

THE HOME OF THE STARS

BOB TOBEN

E. P. DUTTON, INC.　★　NEW YORK

Published in the United States by
E. P. Dutton, Inc., 2 Park Avenue, New York, N.Y. 10016

Library of Congress Catalog Card Number 82-50934
ISBN: 0-525-47629-6

Published simultaneously in Canada by
Clarke, Irwin & Company Limited, Toronto and Vancouver

10 9 8 7 6 5 4 3 2 1

First Edition

HERE IS A TALE OF TALES
OF THE BEGINNINGS

OF STARS
OF THE GALAXIES
OF THE UNIVERSE

OF THE WORLD
AS WE KNOW IT

OF LIGHT
OF THE HEAVENS

HEAVENLY LIGHTS

THE STARS ON A CLEAR NIGHT

GAZING AT THE STARS
ON A CLEAR NIGHT..
WE SEE BUT A TINY GLIMPSE
OF ALL THAT IS.

DEEP WITHIN THE HEAVENS
ARE ALL THE COLORS

THE BEGINNINGS
OF ETERNITY

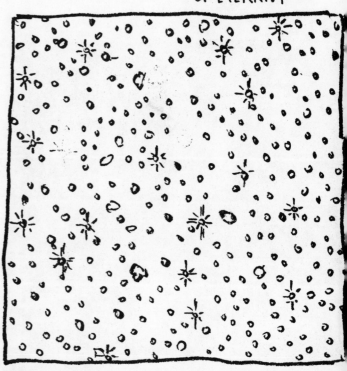

THROUGHOUT HISTORY,
ON THE CLEAREST OF NIGHTS..
ONE MIGHT BE ABLE TO SEE ABOUT
A THOUSAND STARS.

NOW..
WITH TELESCOPES AND CAMERAS..
WE HAVE BEGUN TO SEE
MILLIONS OF STARS AND GALAXIES

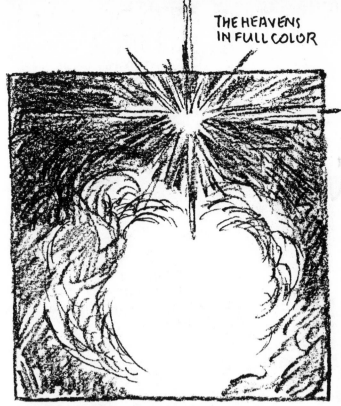

THE HEAVENS
IN FULL COLOR

THE MOST BEAUTIFUL
NRISE

THE MOST BEAUTIFUL
NSET

IN THE GOLDS AND YELLOWS AND
SILVERS AND BLUES AND REDS
AND GREENS AND ORANGES AND
PINKS AND WHITES
OF THE STARS AND CLOUDS OF SPACE

THE MILKY WAY
GALAXY AS SEEN
FROM WITHIN

AND IT'S SAID
THERE ARE BILLIONS OF GALAXIES
IN THE SKIES ..
EACH WITH BILLIONS OF STARS.

WE'VE JUST BEGUN TO SEE
THE HEAVENS IN FULL VIEW ..

WHO KNOWS WHAT WONDERS
AWAIT US ?

HERE'S WHERE WE ARE

THE PLANET EARTH AND THE SUN AS VIEWED FROM SPACE

THE SUN MOVES THROUGH THE STARS NEAR THE OUTER REALM OF THE GALAXY...
AND THE PLANET EARTH MOVES ABOUT THE SUN.

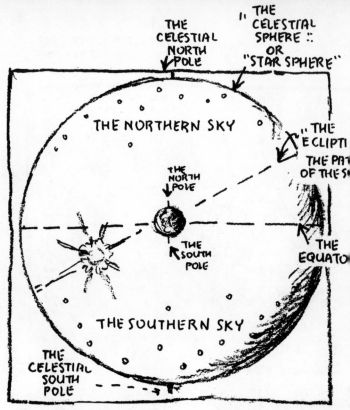

THE CELESTIAL NORTH POLE

THE CELESTIAL SPHERE OR "STAR SPHERE"

THE NORTHERN SKY

THE NORTH POLE

THE SOUTH POLE

THE SOUTHERN SKY

THE CELESTIAL SOUTH POLE

"THE ECLIPTIC"

THE PATH OF THE SUN

THE EQUATOR

TO LOCATE THE STARS AND THE HEAVENLY LIGHTS ALL AROUND US, WE PROJECT AN IMAGINARY "CELESTIAL SPHERE" INTO SPACE . . WITH THE EARTH AT THE CENTER .

ALL THE STARS ARE IMAGINED TO BE MOVING UPON THIS SPHERE . . AS ARE THE SUN AND THE MOON AND ALL THE PLANETS.

THE CELESTIAL NORTH POLE
NEAR THE STAR "POLARIS" IN URSA MINOR

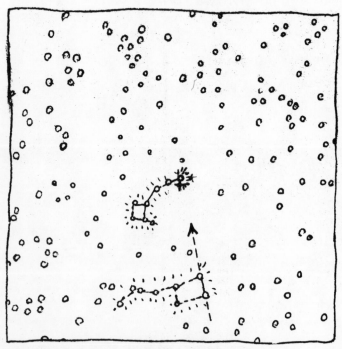

THE "CELESTIAL NORTH POLE" IS DIRECTLY ABOVE OUR GEOGRAPHIC NORTH POLE . .
THE CONSTELLATIONS NEAREST THE CELESTIAL NORTH POLE CAN BE SEEN ONLY FROM THE NORTHERN HEMISPHERE

THE CELESTIAL SOUTH POLE
NEAR THE DIM STAR σ OCTANS" IN OCTANT
AND THE LARGE AND SMALL MAGELLANIC CLOUDS

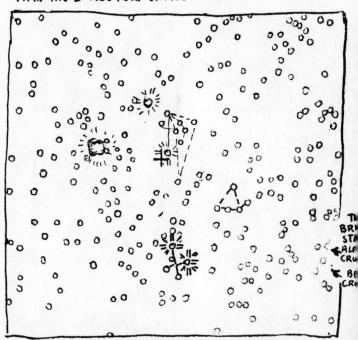

THE BRIGHT STAR ALPHA CRUX

BETA CRUX

THE "CELESTIAL SOUTH POLE" IS DIRECTLY ABOVE OUR GEOGRAPHIC SOUTH POLE . .
THE CONSTELLATIONS NEAREST THE CELESTIAL SOUTH POLE CAN BE SEEN ONLY FROM THE SOUTHERN HEMISPHERE .

THE STARS WE CAN SEE WITHOUT A TELESCOPE VARY QUITE A BIT IN BRIGHTNESS.. THE BRIGHTEST STARS APPEARING ABOUT 100 TIMES BRIGHTER THAN THE FAINTEST STARS.

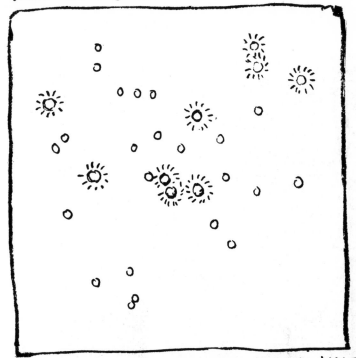

ROUGHOUT STORY..

TARS AVE BEEN ROUPED INTO ONSTELLATIONS"

STARGAZERS OF VARIOUS CIVILIZATIONS THROUGHOUT HISTORY HAVE SEEN DIFFERENT PICTURES IN THE SAME CONSTELLATIONS

OVER THE YEARS, THE CONSTELLATIONS CHANGE A LITTLE IN APPEARANCE TO US ON EARTH, AS THE STARS MOVE IN RELATION TO ONE ANOTHER

WHEN A GALAXY OR OTHER HEAVENLY LIGHT IS SAID TO BE IN A CONSTELLATION.. IT'S NOT WITHIN THE STARS OF THE CONSTELLATION.. BUT IT IS IN THE SAME LINE OF SIGHT AS VIEWED FROM EARTH.. THE STARS OF A CONSTELLATION ARE GENERALLY AT DRAMATICALLY DIFFERENT DISTANCES FROM US.

ITS THOUGHT THAT WE'RE LOCATED IN THE OUTER REGIONS OF THE MILKY WAY GALAXY

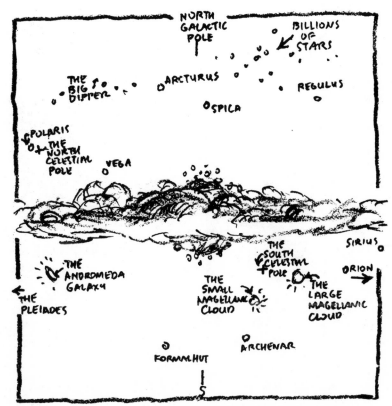

THE MILKY WAY GALAXY IS THOUGHT TO LOOK SOMETHING LIKE THIS, AS VIEWED FROM SPACE.

THE NORTHERN SKY

IN THE EARLY DAYS
OF ASTRONOMY,
THE STARS WERE
VIEWED AS BEING
IN 48 CONSTELLATIONS

ASTRONOMERS OF
TODAY VIEW THE
STARS AS BEING
IN 88 CONSTELLATIONS

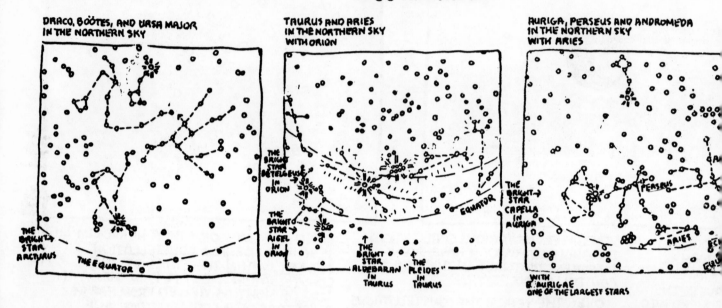

DRACO, BOÖTES, AND URSA MAJOR IN THE NORTHERN SKY

THE BRIGHT STAR ARCTURUS

THE EQUATOR

TAURUS AND ARIES IN THE NORTHERN SKY WITH ORION

THE BRIGHT STAR BETELGEUSE IN ORION

THE BRIGHT STAR RIGEL IN ORION

THE BRIGHT STAR ALDEBARAN IN TAURUS

THE "PLEIDES" IN TAURUS

EQUATOR

AURIGA, PERSEUS AND ANDROMEDA IN THE NORTHERN SKY WITH ARIES

THE BRIGHT STAR CAPELLA IN AURIGA

PERSEUS

ARIES

WITH E. AURIGAE ONE OF THE LARGEST STARS

THE CONSTELLATIONS
NEAR THE CELESTIAL
EQUATOR CAN BE SEEN
FROM MOST LOCATIONS
AROUND THE WORLD

THE SOUTHERN SKY

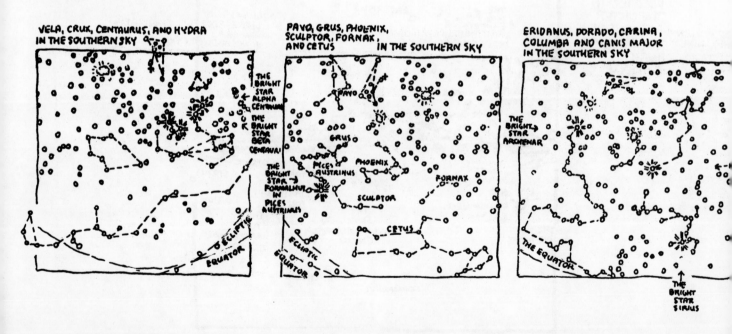

VELA, CRUX, CENTAURUS, AND HYDRA IN THE SOUTHERN SKY

THE BRIGHT STAR ALPHA CENTAURI

THE BRIGHT STAR BETA CENTAURI

THE BRIGHT STAR FORMALHUT IN PICES AUSTRINUS

ECLIPTIC

EQUATOR

PAVO, GRUS, PHOENIX, SCULPTOR, FORNAX, AND CETUS IN THE SOUTHERN SKY

PAVO

GRUS

PICES AUSTRINUS

PHOENIX

SCULPTOR

FORNAX

CETUS

ECLIPTIC OR EQUATOR

ERIDANUS, DORADO, CARINA, COLUMBA AND CANIS MAJOR IN THE SOUTHERN SKY

THE BRIGHT STAR ACHENAR

THE EQUATOR

THE BRIGHT STAR SIRIUS

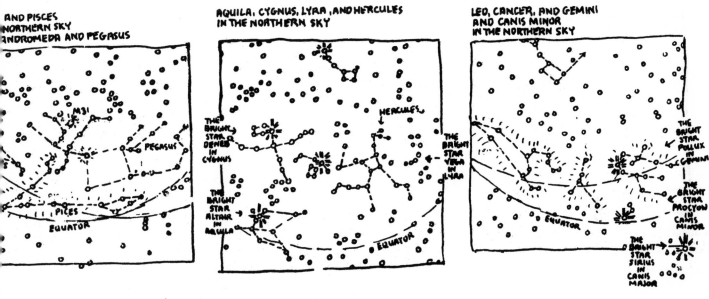

AND PISCES
NORTHERN SKY
ANDROMEDA AND PEGASUS

M31

PEGASUS

PISCES

EQUATOR

AQUILA, CYGNUS, LYRA, AND HERCULES
IN THE NORTHERN SKY

HERCULES

THE BRIGHT STAR DENEB IN CYGNUS

THE BRIGHT STAR ALTAIR IN AQUILA

THE BRIGHT STAR VEGA IN LYRA

EQUATOR

LEO, CANCER, AND GEMINI
AND CANIS MINOR
IN THE NORTHERN SKY

THE BRIGHT STAR POLLUX IN GEMINI

THE BRIGHT STAR PROCYON IN CANIS MINOR

EQUATOR

THE BRIGHT STAR SIRIUS IN CANIS MAJOR

PATH OF THE
UPON THE
ESTIAL SPHERE
ALLED THE
LIPTIC"

THE MOON AND THE
PLANETS APPEAR
TO MOVE THROUGH
THE SKIES
ABOUT IN LINE
WITH THE
"ECLIPTIC"

.. PASSING IN FRONT OF
12 CONSTELLATIONS
OF STARS CALLED
"THE ZODIAC"..
WHICH APPEAR TO
MOVE ALONG THE
SAME PATH

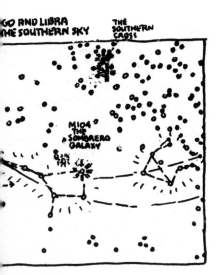

GO AND LIBRA
THE SOUTHERN SKY

THE SOUTHERN CROSS

M104
THE SOMBRERO GALAXY

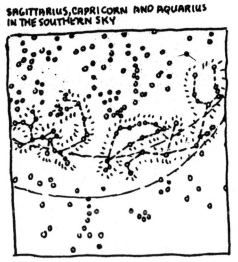

SAGITTARIUS, CAPRICORN AND AQUARIUS
IN THE SOUTHERN SKY

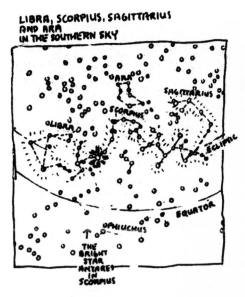

LIBRA, SCORPIUS, SAGITTARIUS
AND ARA
IN THE SOUTHERN SKY

ARA

SAGITTARIUS

SCORPIUS

LIBRA

ECLIPTIC

EQUATOR

OPHIUCHUS

THE BRIGHT STAR ANTARES IN SCORPIUS

THE VISIBLE LIGHT UNIVERSE

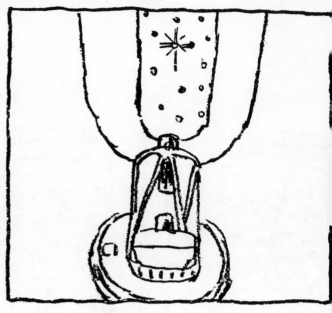

FROM HIGH ATOP THE MOUNTAINS
AROUND THE WORLD

"VISIBLE LIGHT" TELESCOPES ARE
LOOKING TO THE STARS...

GAZING AT STARS
HUNDREDS OF THOUSANDS
TIMES DIMMER THAN
WE NORMALLY SEE

PHOTOGRAPHING LIGHT
FIVE MILLION TIMES DIMMER
THAN WE NORMALLY SEE

THE INVISIBLE UNIVERSES

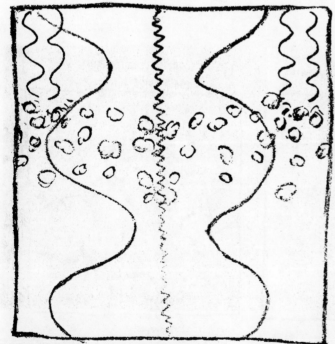

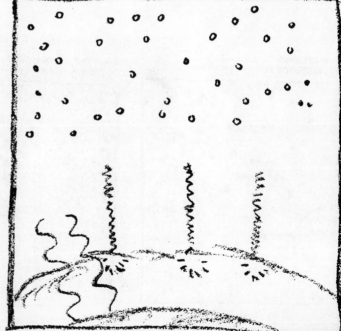

SOME OF THE LIGHT FROM THE STARS
IS BLOCKED BY INTERSTELLAR DUST
AND SOME BY OUR OWN
ATMOSPHERE.
BUT ENOUGH GETS THROUGH
TO GIVE US A DECENT PICTURE

THE HIGHER ENERGY WAVES
ARE SO LITTLE AND INTENSE
THEY PASS RIGHT THROUGH
THE DUST..
BUT ARE ABSORBED BY THE
UPPER ATMOSPHERE OF EARTH

THE LONG LOW ENERGY RADIOWAVES
ARE SO BROAD THEY ARE RELATIVELY
UNAFFECTED BY THE INTERSTELLAR
DUST AND THE ATMOSPHERE OF EARTH

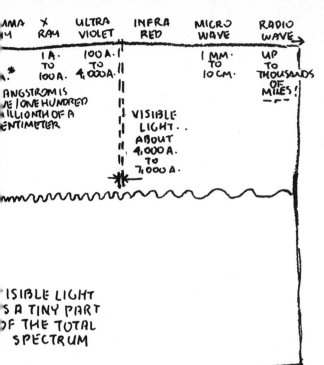

GAMMA | X RAY | ULTRA VIOLET | INFRA RED | MICRO WAVE | RADIO WAVE →

1 A. TO 100 A. | 100 A. TO 4,000 A. | | 1 MM. TO 10 CM. | UP TO THOUSANDS OF MILES!

* AN ANGSTROM IS ONE ONE HUNDRED MILLIONTH OF A CENTIMETER

VISIBLE LIGHT.. ABOUT 4,000 A. TO 7,000 A.

VISIBLE LIGHT IS A TINY PART OF THE TOTAL SPECTRUM

THE LONGEST WAVES ARE ABOUT 1,000,000,000,000,000,000 TIMES AS LONG AS THE LITTLEST WAVES

RADIO WAVES THOUSANDS OF MILES ACROSS.

THE EARTH ..ABOUT 8,000 MILES ACROSS

THE LIGHT THAT WE SEE IS A PERFECT HARMONY OF EVERY WAVELENGTH OF VISIBLE LIGHT

THE VISIBLE LIGHT SPECTRUM IS MADE OF THE COLORS OF THE RAINBOW.. ONE GRADUALLY CHANGING INTO THE NEXT..

STARS SEND OUT A WIDE VARIETY OF WAVES.. TRAVELING AT THE SPEED OF LIGHT.. "THE ELECTROMAGNETIC SPECTRUM"

THE LITTLEST WAVES ARE THOSE OF THE HIGHEST ENERGY.. UP TO MILLIONS OF TIMES MORE ENERGETIC THAN VISIBLE LIGHT WAVES

THE LONGER THE WAVELENGTH.. THE LOWER ITS "ENERGY"..

RADIO WAVES RANGE FROM 1 MM. UP TO THOUSANDS OF MILES ACROSS !

IN THE 1950's BALLOONS AND ROCKETS BEGAN LIFTING TELESCOPES HIGH INTO THE SKY TO GIVE US A LOOK AT THE WAVELENGTHS OF LIGHT BLOCKED BY THE ATMOSPHERE.

TODAY, THERE ARE A GREAT NUMBER OF SATELLITE TELESCOPES ORBITING EARTH.. LOOKING TO THE STARS IN GAMMA, X-RAY, INFRARED, AND VISIBLE WAVELENGTHS.

ON THE SURFACE OF EARTH.. IN THE DESERTS AND PLAINS AND VALLEYS.. GREAT ARRAYS OF RADIO TELESCOPES ARE GIVING US PICTURES IN RADIO WAVE LIGHT.

VISIBLE LIGHT FROM "THE LARGE MAGELLANIC CLOUD"

ULTRAVIOLET LIGHT FROM "THE LARGE MAGELLANIC CLOUD"

SOME OF THE "VISIBLE LIGHT" STARCLOUDS AND GALAXIES LOOK QUITE DIFFERENT WHEN WE SEE THEM IN INVISIBLE LIGHT WAVELENGTHS

SOME STARS AND CLOUDS AND GALAXIES ARE NOT VISIBLE AT ALL IN THE VISIBLE WAVELENGTHS

BUT WE CAN GET A PICTURE OF THEM FROM THE TELESCOPES OF THE "INVISIBLE UNIVERSE".

THE PHOTOGRAPHIC UNIVERSE

LOOKING THROUGH GREAT TELESCOPES,
HUNDREDS OF MILLIONS OF STARS
ARE VISIBLE TO THE EYE.

ON PHOTOGRAPHS, THROUGH THE
SAME TELESCOPES,
BILLIONS OF STARS ARE VISIBLE.

THROUGH TIME PHOTOGRAPHY
WE CAN SEE STARS
HUNDREDS OF MILLIONS OF TIMES
FAINTER THAN THE BRIGHTEST
STARS WE NOW SEE
WITHOUT A TELESCOPE
AND CAMERA.

ONLY A FEW STARS
ARE VISIBLE ON
A CLEAR NIGHT
IN ANY AREA
OF THE SKY..

THROUGH A TELESCOPE,
WE CAN SEE MANY
MORE STARS
IN THE SAME
AREA

BY ADDING A CAMERA
TO THE TELESCOPE,
A GREAT MANY
MORE STARS
APPEAR

AND..IN A TIME
EXPOSURE OF
MORE THAN AN
HOUR..
TENS OF THOUSANDS
OF MORE STARS..
AND EVEN
DISTANT, UNKNOWN
GALAXIES MIGHT
APPEAR!

THE "BIG DIPPER"
PART OF
URSA MAJOR

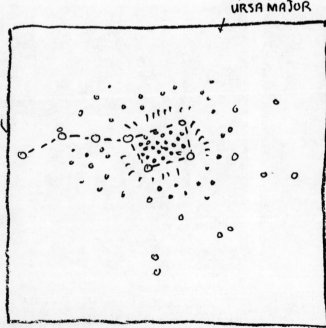

ABOUT
ONE
MILLION
STARS

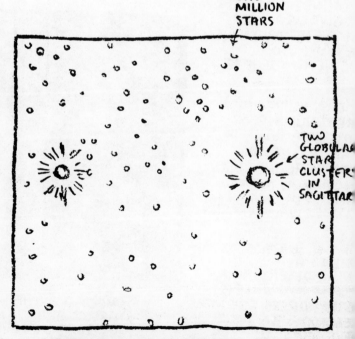

TWO
GLOBULAR
STAR
CLUSTERS
IN
SAGITTARIUS

TODAY WE CAN SEE
ABOUT A MILLION VISIBLE LIGHT GALAXIES
IN THE BOWL OF THE BIG DIPPER ALONE!

TODAY WE CAN SEE ABOUT
A MILLION STARS IN A LITTLE AREA
OF THE SKY AROUND TWO STAR CLUSTERS
NEAR THE CENTER OF THE GALAXY.

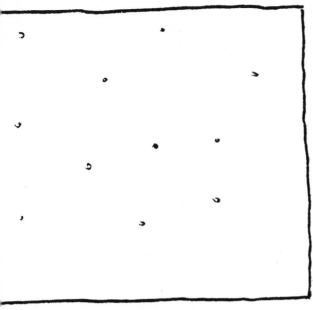

E IMAGES
E SEE TRAVEL
ROSS THE SKIES
LITTLE PHOTONS
LIGHT

THE PHOTONS OF LIGHT
FROM SOME DISTANT
STARCLOUDS, STARS
AND GALAXIES ARE
SO FEW AND FAR
BETWEEN...

THAT EVEN
THROUGH THE
WORLD'S GREATEST
TELESCOPES
WE WOULD SEE
NOTHING.

OVER A LONG
PERIOD OF TIME
THE CAMERA CAN
COLLECT THESE
LITTLE PHOTONS
OF LIGHT ON
PHOTOGRAPHS

PLACING THEM
GENTLY TOGETHER
SO THAT WE MAY
SEE THE WONDERS
OF THE HEAVENS
THROUGHOUT TIME

A GALAXY
OF GALAXIES
↓

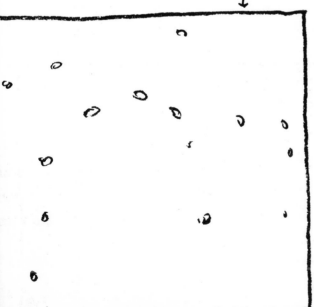

A STAR
FLASHING
11 TIMES
A SECOND
↓

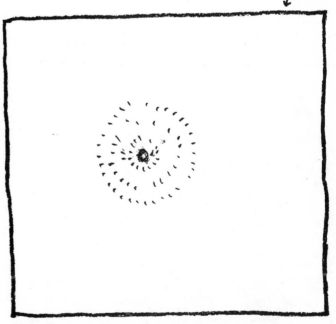

STRONOMERS ARE NOW
ECORDING THE VISUAL LIGHT IMAGES
F OBJECTS
ILLIONS OF LIGHT YEARS AWAY.

THE FAINTEST VISIBLE LIGHT OBJECT WE
NOW SEE IS THE "VELA PULSAR"
A VISIBLE LIGHT/RADIO LIGHT "PULSAR"
FLASHING OPTICALLY 11 TIMES A SECOND.
100,000,000 TIMES TOO FAINT TO BE SEEN
BY THE HUMAN EYE. THE VELA PULSAR
CAN ACTUALLY BE "SEEN" IN PHOTOGRAPHS.

THE UNIVERSE OF ALL TIME

A PHOTON TRAVELING AT THE SPEED OF LIGHT ACROSS THE SOLAR SYSTEM IN ABOUT 12 HOURS →

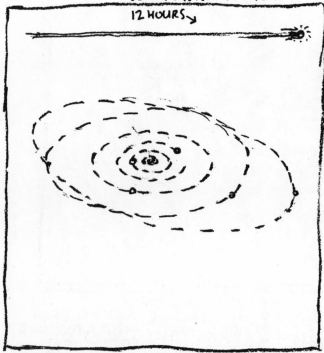

THE DISTANCE THAT A PHOTON OF LIGHT TRAVELS IN ONE SECOND.. 186,282.4 MILES... IS CALLED ONE LIGHT SECOND

A MILLION MILES FROM EARTH.. A LITTLE MORE THAN FIVE LIGHT SECONDS AWAY

THE MOON.. A LITTLE MORE THAN ONE LIGHT SECOND FROM EARTH

THE DISTANCE THAT A PHOTON OF LIGHT TRAVELS IN ONE YEAR.. ABOUT 5.88 TRILLION MILES.. IS CALLED ONE LIGHT YEAR

INSIDE A NEBULA THOUSANDS OF LIGHT YEARS AWAY

WHEN WE LOOK INTO A NEBULA THOUSANDS OF LIGHT YEARS AWAY WE SEE IT AS IT APPEARED THOUSANDS OF YEARS IN THE PAST.

A GALAXY ONE MILLION LIGHT YEARS AWAY

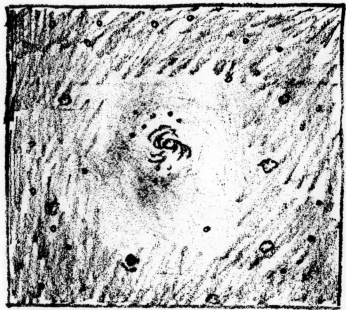

WHEN WE LOOK AT A GALAXY A MILLION LIGHT YEARS AWAY.. WE SEE IT AS IT APPEARED A MILLION YEARS IN THE PAST.

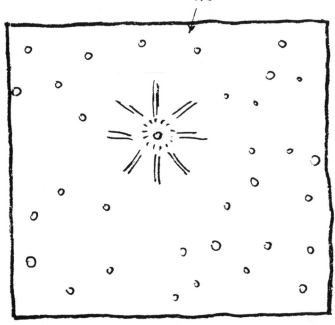

STARLIGHT FROM
A THOUSAND YEARS
AGO

HE SUN IS ABOUT
00 MILLION MILES AWAY

RAVELING AT THE SPEED OF LIGHT
OR ABOUT 100 MILLION MILES,
T TAKES A LITTLE MORE THAN
99 SECONDS FOR A PHOTON OF LIGHT
O ARRIVE ON EARTH.

F THIS IS SO, WE ALWAYS SEE
AN IMAGE OF THE SUN
AS IT WAS ABOUT
99 SECONDS AGO.

THE LIGHT WE SEE FROM A STAR
A THOUSAND LIGHT YEARS AWAY
IS AN IMAGE OF WHAT THAT STAR WAS
A THOUSAND YEARS AGO

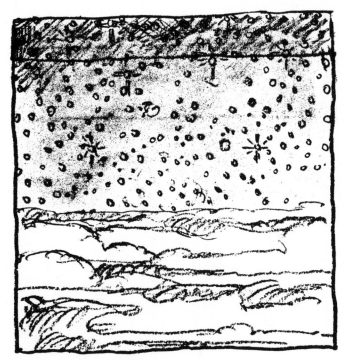

SOME SAY
THE UNIVERSE WE SEE
IS A UNIVERSE OF ALL TIME

THE FURTHER WE LOOK INTO SPACE
THE FURTHER WE LOOK BACK INTO TIME

WE SEE LIGHT FROM "THE PAST"...
WE SEE LIGHT AS IT WAS THE MOMENT
IT BEGAN ITS VOYAGE TO THE EARTH OF
HERE AND NOW.

THE NEAREST STARS

THE CONSTELLATION CENTAURUS

PROXIMA CENTAURI AND ALPHA CENTAURI .. 2 STARS OF A 3 STAR SYSTEM

BARNARDS STAR.. A RED DWARF STAR

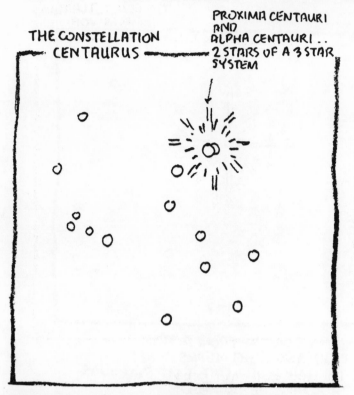

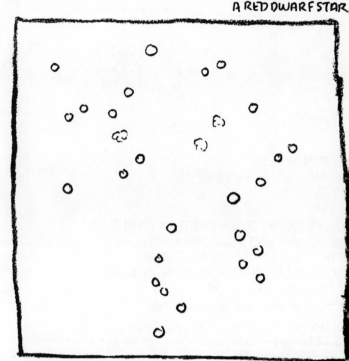

THERE ARE ONLY A FEW STARS WITHIN TEN LIGHT YEARS FROM THE SUN

THE NEAREST STARS ARE PROXIMA CENTAURI!.. NOW ABOUT 4.2 LIGHT YEARS AWAY ... AND ... THE BRILLIANT ALPHA CENTAURI NOW ABOUT 4.3 LIGHT YEARS AWAY

MOST OF THE STARS WITHIN TEN LIGHT YEARS ARE LITTLE STARS

"BARNARD'S STAR" IS A LITTLE STAR ABOUT 5 LIGHT YEARS AWAY.. IT MAY HAVE A PLANET CIRCLING IT IT

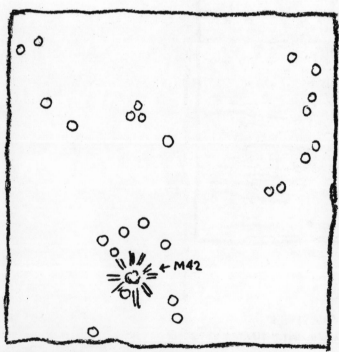

GLOWING IN LUMINOUS COLORS

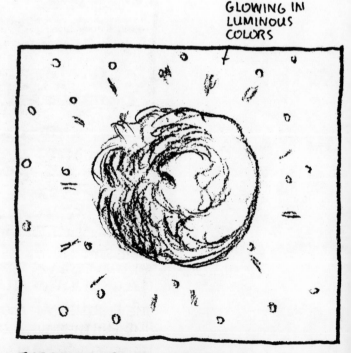

WHAT WAS ONCE THOUGHT TO BE A STAR IN THE MIDDLE OF THE "SWORD" OF ORION IS NOW KNOWN TO BE ONE OF THE NEAREST OF THE GREAT STARCLOUDS

THE GREAT NEBULA IN ORION, M42, ABOUT 1,500 LIGHT YEARS AWAY... WHERE STARS ARE BORN.

SIRIUS "A"
AND
SIRIUS "B"
ITS WHITE DWARF
COMPANION

SIRIUS, THE BRIGHTEST STAR
IN THE SKY, IS ABOUT
9 LIGHT YEARS AWAY

THE BRILLIANT STAR
"PROCYON" IS ABOUT
TEN LIGHT YEARS AWAY

THE MILKY WAY GALAXY
AND HER COMPANIONS

THE NEAREST GALAXIES TO OURS,
ARE "THE LARGE MAGELLANIC CLOUD" AND
"THE SMALL MAGELLANIC CLOUD". -
BOTH LESS THAN 200,000 LIGHT YEARS AWAY

AS FAR AS WE CAN SEE

ASTRONOMERS DETERMINE DISTANCES OF THE NEAREST GALAXIES BY OBSERVING THE MOVEMENTS OF BRILLIANT PULSATING VARIABLE STARS WITHIN THOSE GALAXIES

IN THE FURTHER GALAXIES ..
THE BRIGHTEST STARS. . .
THE BRIGHTEST NOVAS. . .
THE BRIGHTEST CLUSTERS OF STARS
WITHIN THOSE GALAXIES

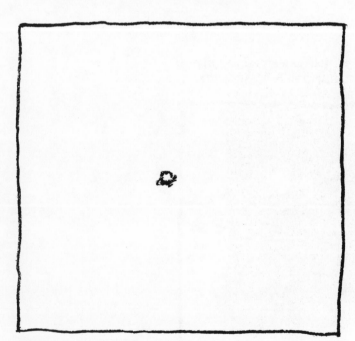

"VISIBLE LIGHT" GALAXIES HAVE BEEN FOUND FIVE BILLION LIGHT YEARS AWAY . .

AND "RADIO GALAXIES" HAVE BEEN DISCOVERED EVEN FURTHER AWAY

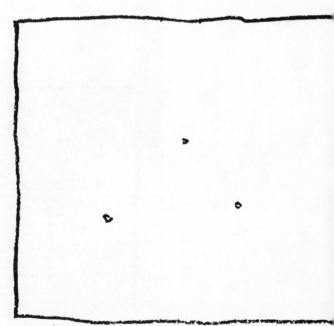

QUITE RECENTLY, 3 GALAXIES * HAVE BEEN DISCOVERED THAT APPEAR TO BE ABOUT TEN BILLION LIGHT YEARS FROM EARTH .

THE LIGHT FROM THESE GALAXIES IS ABOUT 50 TIMES DIMMER THAN THE NORMAL "GLOW" OF THE NIGHT TIME SKIES!

* BOTH OPTICAL AND RADIO RADIATION

IN THE MOST DISTANT
REALMS OF SPACE
AS WE KNOW IT ···
EVEN WITH THE MOST
POWERFUL TELESCOPES
OF TODAY ··
THE DISTANCES TO
FAR FAR AWAY GALAXIES
AND QUASARS
ARE VERY DIFFICULT
TO DETERMINE .

AND, IN THE FURTHER REACHES
OF THE VISIBLE LIGHT "UNIVERSE
·· THE BRIGHTEST GALAXIES
WITHIN GALAXIES OF GALAXIES

IN THE MOST DISTANT REACHES··
DISTANCE JUDGEMENTS
ARE NOT MUCH MORE
THAN
EDUCATED GUESSES .

RECENTLY,
QUASARS HAVE BEEN
OBSERVED
THAT ARE THOUGHT
TO BE 15 BILLION
LIGHT YEARS
AWAY.

SOON WE SHOULD BE
ABLE TO "SEE"
TO THE LIGHTS
AND STARS
AND GALAXIES
12 BILLION
LIGHT YEARS
AWAY.

AND ·· IN THE
NOT TOO DISTANT FUTURE ··
TO THE LIGHTS
AND STARS
AND GALAXIES
15 TO 17 BILLION
LIGHT YEARS
AWAY.

THE SPACE TELESCO[PE]
SENDING BACK
LITTLE MOVIES
OF FAR AWAY
PLACES

THE "SPACE TELESCOPE"
IS SCHEDULED TO BE
LAUNCHED IN 1985 ...
TO ORBIT ABOVE
THE UPPER ATMOSPHERE

SOME SAY THE
DETAIL OF THE PHOTOS
WILL BE TEN TIMES
AS CLEAR AS WE
GET NOW

AND
WE'LL BE ABLE
TO SEE ABOUT
A THOUSAND
TIMES MORE
SPACE !

WE SHOULD BE ABLE T[O]
SEE LIGHTS 100 TIME[S]
DIMMER THAN WE NO[W]
SEE WITH OUR BEST
MOUNTAIN TOP
TELESCOPES .

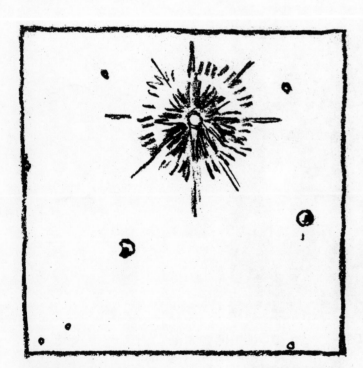

WE SHOULD BE ABLE TO SEE
PLANETS
CIRCLING STARS
FAR AWAY

AND.. SOMEDAY..
ZOOM IN
FOR A
CLOSE-UP
VIEW

SOON
WE SHOULD BE
ABLE TO SEE
THE CENTER OF THE GALAXY

IN FULL COLOR

SOON
WE SHOULD BE
ABLE TO SEE
DEEP WITHIN THE CLOUDS
WHERE THE STARS
COME FROM

IN FULL COLOR

WHO KNOWS
WHAT WONDERS
ARE WAITING
?

STARS

STARCLOUDS

NEBULAE ARE
THE BEGINNINGS OF THE STARS
AND
THE ENDINGS OF THE STARS

IN THE GREAT NEBULA IN
THE "SWORD" OF ORION ARE
FOUR STARS ..
ABOUT 1,000 TIMES
MORE BRILLIANT
THAN THE SUN

THEY
LIGHT UP THE SKY
IN A REGION
100 TIMES BIGGER
THAN OUR
SOLAR SYSTEM

DIFFUSE NEBULAE ARE THE CLOUDS
WHEREIN THE STARS ARE BORN ...
BILLOWING CLOUDS OF COSMIC GAS
AND STARDUST ..
THEY OFTEN GLOW FROM THE LIGHT
OF THE YOUNG STARS WITHIN.

THE GREAT NEBULA IN ORION IS THE BRIGHTEST
IN THE SKY ... THE HOME OF THE BLUEST AND
MOST MASSIVE YOUNG STARS IN THIS GALAXY.

WITHIN
THE LAGOON NEBULA
IN SAGITTARIUS
6,500 LIGHT YEARS
AWAY

VARIABLE STARS
SOMETIMES
FLARE UP TO 25 TIMES
BRIGHTER THAN
NORMAL

IN THE "ROSETTE" NEBULA
THERE ARE DARK SPHERES ..
STARS NOT YET BORN ? ..
TENS OF THOUSANDS OF TIMES
BIGGER THAN THE SUN.

THROUGHOUT
THE HEAVENS
OF ALL TIME

THE CLOUDS OF THE STARS
ARE A GLOW
IN AMAZINGLY
BEAUTIFUL
LIGHT

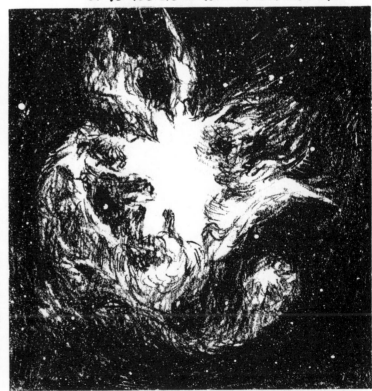

THE TRIFID NEBULA, M20, [I]N "EMISSION" NEBULAE, DIFFUSE NEBULAE CONTAIN
[IN] SAGITTARIUS. ABOUT SOME STARS EXCITE THE STRONG INFRA RED LIGHT
[30]OO LIGHT YEARS AWAY. HYDROGEN GAS INTO A SOURCES. WHICH COULD BE
[AN] EXPANDING, GLOWING, SELF-LUMINOSITY GREAT "PROTOSTARS"
[DI]FFUSE "EMISSION" NEBULA.

DEEP WITHIN THE DORADO NEBULA.
A CLUSTER OF BLUE SUPERGIANT STARS
EXCITES THE HYDROGEN TO ILLUMINATION
UP TO 400 LIGHT YEARS FROM THE STARS

THE
ROSETTE
NEBULA

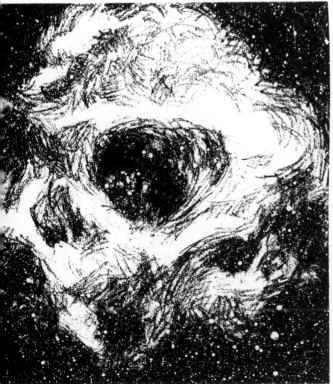

THE DORADO NEBULA   -- A DIFFUSE NEBULA
IS THE BRIGHTEST   AS LUMINOUS AS
NEBULA IN THE   50,000 SUNS!!
"LARGE MAGELLANIC CLOUD"
GALAXY

SOME NEBULAE ARE DARK
SOME REFLECT LIGHT
SOME GLOW

"DARK" NEBULAE
ARE COSMIC CLOUDS
WITH NO STARS WITHIN
TO LIGHT THEM UP.

THE DARK, EMPTY LOOKING
REGIONS OF THE MILKY WAY
ARE CLOUDS
OF DARK NEBULAE
COVERING THE LIGHT OF
DISTANT STARS BEYOND

THE "COALSACK" NEBULA
NEAR THE SOUTHERN CROSS
IS A DARK NEBULA.

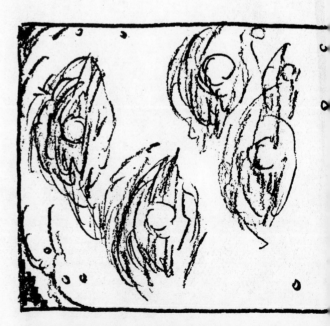

"REFLECTION" NEBULAE
ARE INTERSTELLAR DUST
LIT BY REFLECTED LIGHT
FROM NEARBY STARS

THE INTERSTELLAR DUST
ILLUMINATED BY THE REFLECTED LIGHT
OF THE STARS OF THE PLEIDES
IS A "REFLECTION" NEBULA

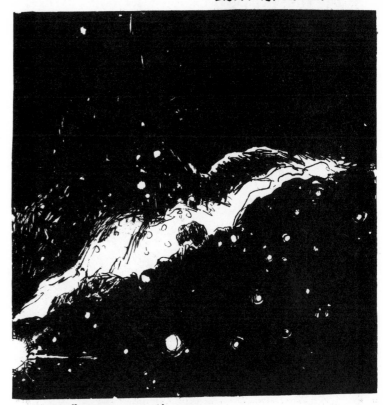

BARNARD'S "S" NEBULA IN THE MILKY WAY . .
N THE CONSTELLATION OPHIUCHUS . .
IS A DARK NEBULA . .

THE "HORSEHEAD" NEBULA . .
NEAR THE SOUTHERNMOST STAR IN
THE BELT OF ORION . .
IS A DARK NEBULA IN FRONT OF
A LUMINOUS ONE

APPROACHING
THE CONE NEBULA
IN MONOCEROS

SOME STARCLOUDS ARE
COMBINATIONS OF
DIFFERENT TYPES
OF NEBULAE

THE TOP OF THE CONE NEBULA
IN MONOCEROS IS THOUGHT TO
GLOW FROM THE ENERGY OF
A VERY BRIGHT STAR ABOVE IT.

SOME NEBULAE ARE THE ENDINGS OF STARS
SOME ARE THE DUST OF A TRANSFORMATION

THE CRAB NEBULA IN TAURUS IS
THE REMAINS OF A SUPERNOVA EXPLOSION...
GETTING BIGGER AT 1,000 MILES A SECOND

AT THE CENTER OF WHICH
IS A STAR IN ONE OF ITS LAST
STAGES..
A RAPIDLY SPINNING LITTLE
NEUTRON STAR..OR "PULSAR"

A "PLANETARY NEBULA"
IS A GLOWING SHELL OF
EXPANDING DUST FROM
THE TRANSFORMATION
OF A STAR
THAT TOOK PLACE
MANY YEARS AGO

PLANETARY NEBULAE
HAVE NOTHING TO DO
WITH PLANETS..
THEY'RE CALLED
"PLANETARY" BECAUSE
A LOT OF THEM LOOK
LIKE SPHERES

THE
"DUMBELL"
NEBULA

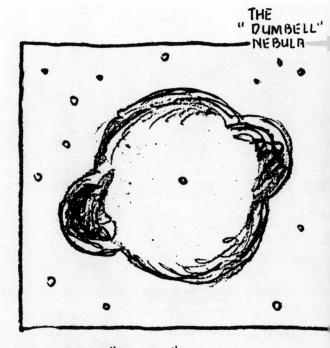

THE NEARBY "DUMBELL"
NEBULA IN VELPECULA
IS A PLANETARY
NEBULA..SURROUNDING
A NOT-SO-BRIGHT STAR..
ABOUT 700 LIGHT
YEARS AWAY

THE
VEIL
NEBULA

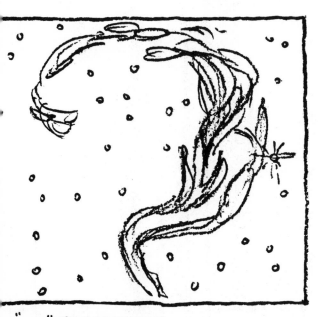

HE "VEIL" NEBULA IN CYGNUS
BOUT 1,000 LIGHT YEARS AWAY
S AN ENDING · ·
LOWING, COLORED GASES · ·

THE LAST WISPS FROM A
SUPERNOVA EXPLOSION
ABOUT 20,000 YEARS AGO

THE
"RING"
NEBULA

THE   THE
"GUM"   REMAINS
NEBULA   OF A
             SUPERNOVA ?

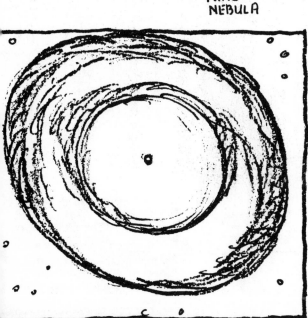

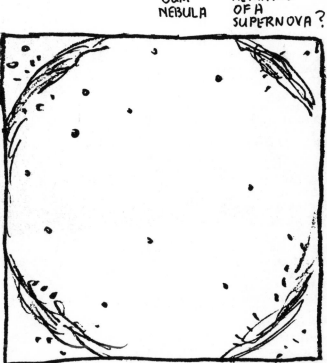

HE "RING" NEBULA IN LYRA · ·
BOUT 5,000 LIGHT YEARS AWAY · ·
S A PLANETARY NEBULA · ·
RILLIANTLY COLORED
LOWING GAS AROUND A
RIGHT CENTRAL STAR

THE VERY FAINT APPEARING
"GUM NEBULA" IN THE
SOUTHERN SKIES · · THOUGHT TO BE
ABOUT 1,200 LIGHT YEARS ACROSS · ·
IS ABOUT 160 TIMES AS BIG AS
THE GREAT NEBULA IN ORION

A GREAT CLOUD OF FLOURESCENT GAS · ·
THE BIGGEST INTERSTELLAR "BUBBLE"
KNOWN · · STILL EXPANDING AND
GLOWING FROM, PERHAPS,
A SUPERNOVA ABOUT
20,000 YEARS AGO

INTERSTELLAR CLOUDS

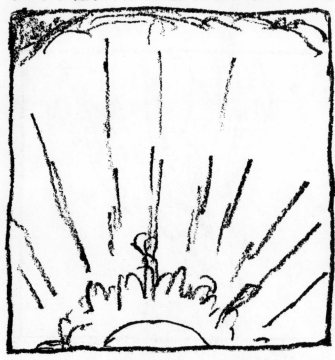

FORMED BY THE STELLAR WINDS

AND SUPERNOVA EXPLOSIONS

AND ENERGY FROM SOURCES UNKNOWN

COSMIC GAS CLOUDS SOMETIMES TRAVEL AMONG THE STARS AS GIGANTIC GLOWING BALLOONS

THE "INTERSTELLAR MEDIUM" IS MADE OF GAS AND DUST · · IT IS THROUGHOUT SPACE · · IN THE GALACTIC PLANE IT'S MORE CONCENTRATED · · HEAVIEST TOWARD THE CENTER ·

THE "INTERSTELLAR GAS" IS MOSTLY INDIVIDUAL ATOMS OF HYDROGEN · WITH SOME ATOMS OF HELIUM · · AND CARBON, NITROGEN, OXYGEN, NEON, IRON, SULFUR, SILICON, ETC · · AND SOME MOLECULES OF HYDROGEN AND MORE ·

RECENTLY · · SOME X-RAY ASTRONOMERS HAVE OBSERVED HUGE CLOUDS OF SUPERHOT PRIMORDIAL INTERGALACTIC GAS · · OF HYDROGEN AND OF HELIUM · · THAT ALSO CONTAIN AMMONIA, FORMALDEHYDE, AND WATER ·

THE INTERSTELLAR DUST IS MOSTLY PARTICLES, OR "GRAINS" OF A MILLION OR MORE ATOMS EACH · THE DUST IN SPACE IS MOSTLY GRAINS OF CARBON ATOMS IN LONG CHAINS (GRAPHITE!) AND GRAINS OF COMPOUNDS OF HYDROGEN · ·

THE AVERAGE DUST GRAIN IS ONE TEN THOUSANDTH OF A CENTIMETER ACROSS · · WITH A COATING OF ICE A FEW ATOMS THICK

GREAT SPHERES OF INTERSTELLAR CLOUDS ROAM THE HEAVENS

"INTERSTELLAR CLOUDS" MOVE ABOUT THE CENTER OF OUR GALAXY IN THE GALACTIC PLANE · · AS DOES THE INTERSTELLAR GAS ·

THE INTERSTELLAR CLOUDS ARE TEN TO A THOUSAND TIMES AS DENSE AS THE INTERGALACTIC GAS ·

THE INTERSTELLAR CLOUDS ARE ATOMS OF HYDROGEN AND HELIUM, CARBON AND OXYGEN AND MOLECULES OF CARBON MONOXIDE FORMALDEHYDE, METHYL ALCOHOL AND MORE · ·

RECENTLY · SOME ASTRONOMERS HAVE OBSERVED AMINO ACID MOLECULES IN THE CLOUDS · · THE ESSENTIAL "LIFE" MOLECULE OF ALL LIVING THINGS OF EARTH

TERSTELLAR GAS
BITS THE CENTER
SPIRAL GALAXIES
THE GALACTIC PLANE

..IT'S ABOUT TEN TIMES
MORE DENSE IN THE
SPIRAL ARMS OF
OUR GALAXY
THAN
IN BETWEEN

INTERSTELLAR GAS
IS MILLIONS OF TIMES
THINNER THAN THE
BEST VACUUM
THAT CAN BE MADE
ON EARTH

IN THE "EAGLE NEBULA"
IN THE CONSTELLATION SERPENS
WE CAN SEE SOME
BEAUTIFUL FORMATIONS
CAUSED BY THE MEETINGS
OF DIFFERENTLY IONIZED
ATOMS OF GAS.

OVER BILLIONS OF YEARS
OF STARS LIGHTING
AND TRANSFORMING
AND EXPLODING
AND SENDING
FORTH ATOMS

THE CLOUDS
MAY GATHER AND CONTAIN
ALL THE ELEMENTS
THAT WE ARE MADE OF
AND THAT SUSTAIN LIFE
HERE ON EARTH.

SOME SAY THAT
WE ARE MADE OF
THE SAME ATOMS
THAT WERE HERE
SINCE
THE BEGINNING OF
THIS PLANET.

.. ATOMS
FROM THE STARS..
FROM THE SKIES
THOUSANDS OF
LIGHT YEARS
AWAY.

THE STARS

THE SUN
IS A STAR ON ITS OWN..
NOT ONE OF A CLUSTER
NOR OF ANY
MULTIPLE STAR
SYSTEM.

ALTHOUGH IT MAY
HAVE BEEN BORN
AT ABOUT THE SAME TIME..
IN THE SAME CLOUD
WITH OTHER NEARBY
STARS..
IT MOVES ALONE

MORE THAN HALF OF
THE STARS OF OUR GALAXY
ARE BORN IN GROUPS OF
TWO OR MORE

MOST STARS ARE
BORN IN GROUPS..
OF THE SAME AGE
MOVING IN THE
SAME INITIAL MOTIO
AS THE CLOUD.

AN OPEN STAR CLUSTER IS A LOOSE GROUPING
OF UP TO THOUSANDS OF STARS THAT MOVE
TOGETHER THROUGH SPACE

THEY ARE GENERALLY FOUND NEAR
THE PLANE OF THE MILKY WAY.

THERE ARE LOTS OF "OPEN STAR CLUSTERS"
IN THE MILKY WAY GALAXY,
BY SOME ESTIMATES, UP TO 100,000 OF THEM.

SOME OPEN STAR CLUSTERS.
LIKE THE "SEVEN SISTERS" IN THE PLEIDES..
410 LIGHT YEARS AWAY..
CAN BE SEEN FROM EARTH...

OPEN STAR CLUSTERS MORE THAN
20,000 LIGHT YEARS AWAY ARE HIDDEN
FROM OUR VIEW BY THE COSMIC DUST.

AN
OPEN
STAR CLUST

SOME OPEN STAR CLUSTERS HAVE BEEN
AROUND A LONG TIME..
LIKE THE OPEN STAR CLUSTER IN CANCER..
WHICH IS ONE OF THE OLDEST IN
THE MILKY WAY GALAXY..

SOME ARE RELATIVELY YOUNG.

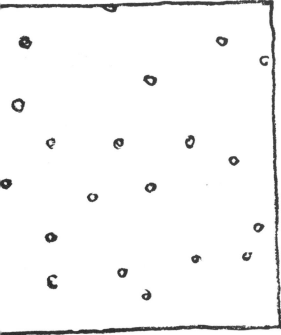

STELLAR ASSOCIATIONS

ELLAR ASSOCIATIONS
E STARS THAT HAVE
COMMON ORIGIN
D ARE
ITE SIMILAR

BUT ARE FAR APART
--AND NOT BOUND
GRAVITATIONALLY
TO EACH OTHER

STARS WITHIN THE
SPIRAL ARMS OF
THE MILKY WAY
ARE NORMALLY IN
"STELLAR ASSOCIATIONS"

THE GLOBULAR
STAR CLUSTER M22
IN SAGITTARIUS..
THE FIRST DISCOVERED

GLOBULAR
STAR CLUSTERS
IN THE
GALACTIC
HALO

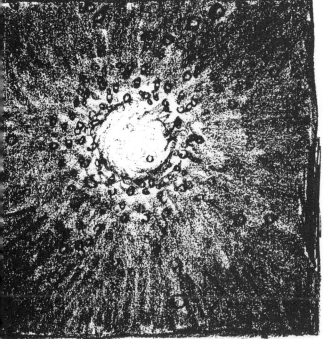

NDREDS OF THOUSANDS
STARS
GREAT, DAZZING
HERICAL GROUPS
E CALLED
OBULAR
AR CLUSTERS"

THEY'RE THOUGHT TO BE
THE OLDEST GROUPINGS
OF STARS IN THE GALAXY
..A TYPICAL ONE CONTAINS
500,000 STARS!

THEY POPULATE THE
GALACTIC HALO..
A REGION
SURROUNDING
THE CENTRAL "DOMES"

IN THE MILKY WAY GALAXY..
THERE ARE ABOUT 200
GLOBULAR STAR CLUSTERS..
THEY'RE SO BRIGHT
WE CANT SEE THE
INDIVIDUAL STARS.

THROUGHOUT ALL OF TIME
THE WAVES FROM NEW STARS
AND OLD STARS
AND EXPLODING STARS

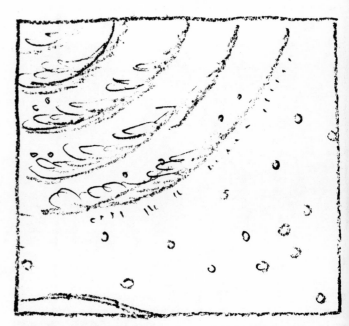

TRAVEL THE GALAXY
TO JOIN TOGETHER INTO
GREAT STARWAVES

SOME CLOUDS
MOVE INWARD
WITHIN
THEMSELVES

AND, AS
WAVES FROM
ACROSS THE SKIES
MOVE THROUGH
THE CLOUDS,

THE CLOUDS BILLOW
IN THE WAKE OF THE WAVES

AND CLOUDS WITHIN THE
CLOUDS
COME TO BE

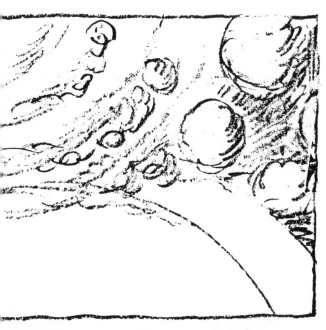

THE GREAT STARWAVES
CARRY THE DUST OF
THE STARS
WITH THEM

WHERE THE WAVES MEET,
CLOUDS APPEAR
OUT OF THE DUST OF
THE STARS
FROM ACROSS THE HEAVENS

SO THAT NOW
AND EVERY NOW AND THEN
HERE AND THERE
THROUGHOUT THE SKIES
GREAT NEW CLOUDS APPEAR.

AND IN THE CLOUDS
WITHIN THE CLOUDS
THE STARS BEGIN

NO ONE KNOWS
HOW THE PLANETS BEGAN
SOMEDAY
WE MAY SEE
THE BEGINNING OF
ANOTHER SOLAR SYSTEM

UNTIL THEN...
HERE IS SCIENCE'S
MOST POPULAR STORY
OF TODAY
OF THE BEGINNING
OF THE PLANETS

A GREAT CLOUD.
TEN THOUSAND
TIMES BIGGER
THAN THE
SOLAR SYSTEM
OF TODAY

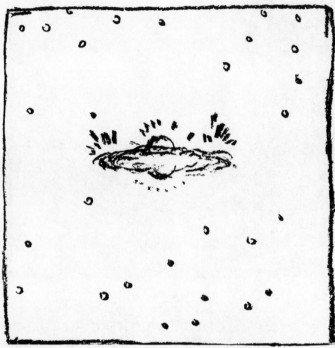

IT'S THOUGHT THAT
OUR SOLAR SYSTEM BEGAN
ABOUT 5 BILLION YEARS
AGO...

THE GALAXY
WAS ALREADY
VERY VERY OLD

AND GENERATIONS OF STARS
HAD COME AND GONE

WITHIN THE ARMS OF
THE GALAXY,
THERE WAS A GREAT
CLOUD...

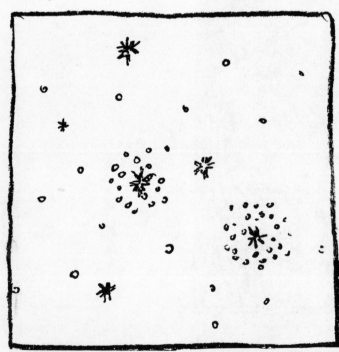

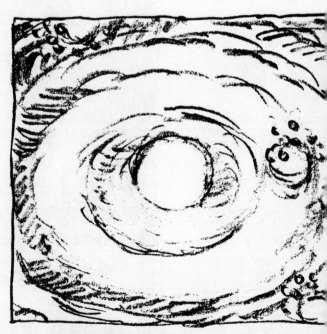

WITHIN THE CLOUD...
LITTLE CRYSTALS OF ICE
AND THE DUST OF STARS
HAD COME TOGETHER
INTO LITTLE "SPHERES"
OF SNOW

THAT, OVER THOUSANDS
OF YEARS, GREW
BIGGER AND BIGGER
AND JOINED
TOGETHER

AND BEGAN REVOLVING
IN CLOUD-LIKE RINGS
WITHIN A GREAT
DISC-SHAPED CLOUD

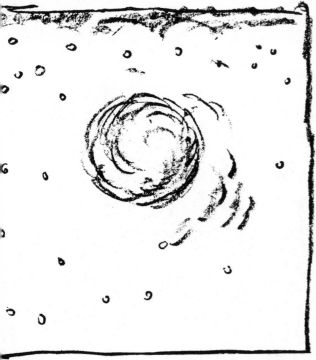

OUCHED
Y GREAT STARWAVES
ROM
CROSS THE SKIES

THE CLOUD
BEGAN TO REVOLVE
AND
MOVE INWARD

AND CLOUDS
WITHIN THE CLOUDS
BEGAN

WHAT WOULD
SOMEDAY BECOME
THE SUN
AND THE PLANETS

THE
PROTOPLANETS
BEGIN TO
APPEAR

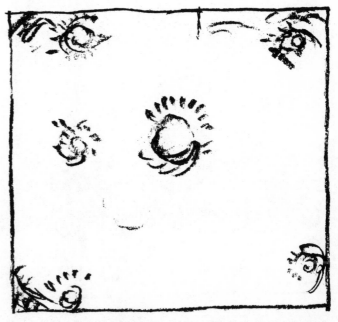

COME TO BE
REAT REVOLVING
HERE CLOUDS
F GAS AND DUST

MOVING INWARD
CIRCLING
AND CENTERED ABOUT
A NEW PROTOSTAR.

THE SPHERES BEGAN TO GLOW

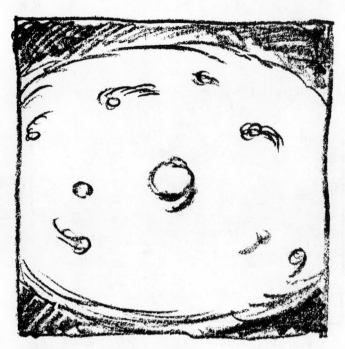

AFTER MILLIONS OF YEARS,
THE YOUNG SOLAR SYSTEM
APPEARED

ZOOM IN TOWARD
THE CENTER

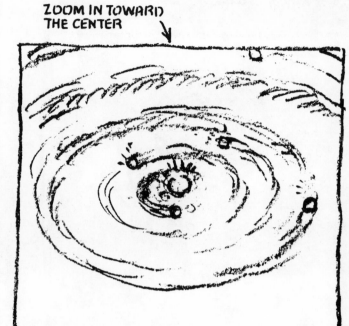

AS THE GREAT CLOUD
MOVED INWARD,
THE PROTOPLANETS
NEAREST THE STAR
SPINNED THE FASTEST...

MOVING FURTHER
INWARD WITHIN
THEMSELVES...
WITHIN THEIR
"CLOUDS"

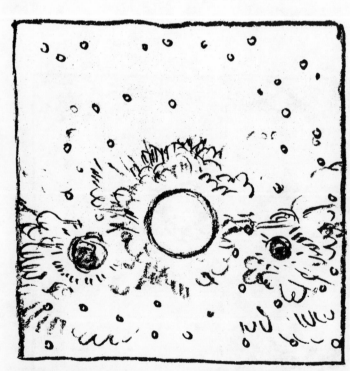

THE WAVES OF
THE LIGHTING OF THE SUN
SOARED OUTWARD
IN ALL DIRECTIONS

BLOWING
THE BLANKETS OF CLOUDS
AWAY FROM
THE
INNER PLANETS

BUT THE WAVES OF
THE LIGHTING OF THE SUN
LOST STRENGTH IN
THE OUTER REACHES...
IN THE REALM OF
THE OUTER PLANETS

-STILL STRONG ENOUGH TO,
PERHAPS, UNCOVER PLANETARY
"RINGS" AND MOONS, BUT
NOT STRONG ENOUGH TO
UNCOVER THE "BLANKETS"
OF THE OUTER PLANETS.

ZOOM OUT
TO THE OUTER REGIONS →

THE OUTER REGIONS
UR GIANT SPHERES
RCLED AND MOVED
WARD..

AS THE SYSTEM
COOLED, ICES
BEGAN FORMING
ON THE GIANT
OUTER SPHERES

AND THEY GREW
EVEN LARGER..
ATTRACTING
GAS AND DUST
AND DISCS.

ONE DAY,
THE SUN
FLASHED
FORTH
AS A
STAR

AND TO THIS DAY.
THE GIANT PLANETS
STILL RIDE THE SKIES
NOT QUITE PLANETS,
NOT QUITE STARS...

LOOKING PRETTY MUCH
AS THEY DID
BEFORE THE
LIGHTING OF
THE SUN

THE PLANETS

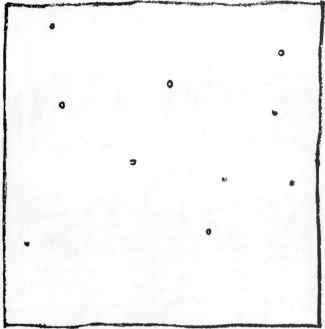

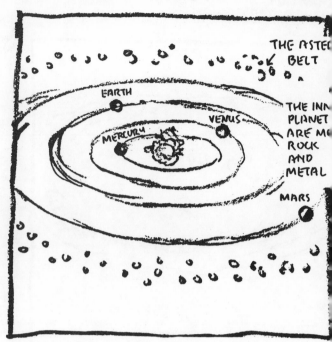

THE ASTEROID BELT

EARTH

VENUS

MERCURY

MARS

THE INNER PLANETS ARE MOSTLY ROCK AND METAL

THIS STAR, THE SUN
WITH ITS NINE PLANETS AND THEIR MOONS
IS CALLED "THE SOLAR SYSTEM"

SEEN IN THE NIGHT SKY FROM EARTH,
THE PLANETS LOOK LIKE BRILLIANT "STARS",
REFLECTING LIGHT FROM THE SUN

THE SOLID SURFACED "INNER PLANETS"
ORBIT CLOSEST TO THE SUN..
MERCURY, VENUS, EARTH AND MARS

THE SAME SIDE
OF THE MOON
ALWAYS
FACES EARTH..

MERCURY VENUS EARTH MARS THE ASTEROIDS

THE EARTH AND THE MOON ROTATE ABOUT THE SUN AS A "DOUBLE PLANET" SYSTEM

THE MOON IS MADE OF ROCK ..

VERY MUCH LIKE THE EARTH

IT'S THOUGHT THAT THE MOON WAS FORMED AT THE SAME TIME AS THE EARTH. BUT VERY LITTLE IS KNOWN ABOUT HOW THIS MIGHT HAVE COME TO BE.

THE PLANETS ORBIT THE SUN AT VERY GREAT DISTANCES APART FROM EACH OTHER.
IF WE COULD SEE THEM CLOSE TOGETHER, THEY'D LOOK SOMETHING LIKE THIS...

BEYOND MARS ARE THOUSANDS
OF LITTLE PLANETS..
THE ASTEROIDS

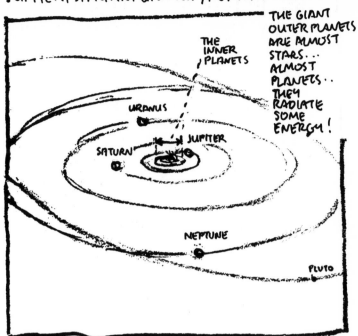

THE OUTER PLANETS...
JUPITER, SATURN, URANUS, NEPTUNE AND PLUTO

THE GIANT
OUTER PLANETS
ARE ALMOST
STARS...
ALMOST
PLANETS..
THEY
RADIATE
SOME
ENERGY!

THE
INNER
PLANETS

URANUS

JUPITER

SATURN

NEPTUNE

PLUTO

FAR BEYOND THE
ASTEROIDS, THE
GREAT "GAS GIANTS"..
JUPITER,
SATURN,
URANUS,
NEPTUNE..
ORBIT THE SUN.

IN THE OUTER
REACHES OF
THE SOLAR
SYSTEM..
IS THE TINY
PLANET
"PLUTO"

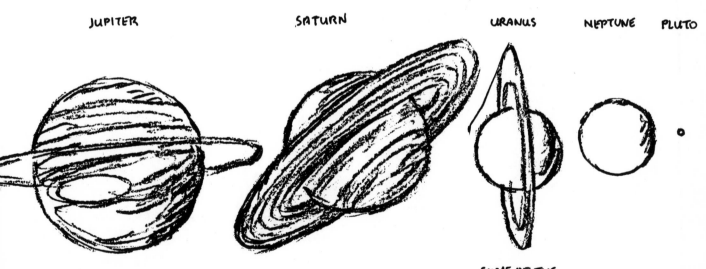

JUPITER   SATURN   URANUS   NEPTUNE   PLUTO

SOME OF THE
PLANETS HAVE
A LOT OF
"MOONS"

A NUMBER OF
SPACE-TRAVELING
"PLANETARY PROBES"
HAVE BEGUN TO
GIVE US OUR
FIRST GOOD
"CLOSE-UP" LOOK
AT THE PLANETS

TRAVELING
AS FAR AS
SATURN
SO FAR.

ONE OF THE
MOONS OF
SATURN..
"TITAN"
IS BIGGER
THAN
MERCURY

.. AND
ONE OF THE
MOONS OF
NEPTUNE..
"TRITON"..
IS BIGGER
THAN
MERCURY

THE PLANET EARTH

WATER
COVERS MOST OF
THE SURFACE
OF THE PLANET

THIS PLANET EARTH
IS THE
TREASURE OF TREASURES
OF THE SOLAR SYSTEM

THERE'S WATER
WITHIN EARTH
AND ON EARTH

IN THE OCEANS AND
THE SEAS

AND RIVERS
AND STREAMS

IN THE AIR
AND
IN THE CLOUDS
AND
IN SPACE

TOGETHER
THE SUN AND THE EARTH
AND THE WATERS
CREATE THE CLIMATES
AND THE WEATHER
OF THE WORLD

STORMS
SOMETIMES ARISE
OVER THE WATERS
BECAUSE OF THE
WARMING FROM THE SUN
AND THE ROTATION
OF THE EARTH

THE REGION OF THE EQUATOR
RECEIVES MORE LIGHT
THAN THE
NORTH AND SOUTH "POLES"

WARMING THE AIR
WHICH RISES
AND BEGINS TO FLOW
TO THE NORTH AND SOUTH

WHERE IT COOLS
AND BEGINS ITS DESCENT
BACK TO EARTH
AND ITS RETURN
TO THE EQUATOR

WITHIN THE
GREAT OCEANS OF EARTH
MOVING LIKE "RIVERS"
WARM STREAMS
FLOW "UP" TO WARM
THE COOLER LANDS
OF THE NORTH AND SOUTH

COOL STREAMS
FLOW "DOWN" FROM
THE ARCTIC AND ANTARC
BACK TOWARD
THE EQUATOR

THE PLANET EARTH
IS MADE OF 7 VARIETIES
OF CRYSTAL
IN A NUMBER OF
COMBINATIONS

WATER
IS LIQUID CRYSTAL
SHIMMERING...
EVERCHANGING...

IN THE HIGHEST OF CLOUDS
CRYSTALS OF ICE
APPEAR DIRECTLY
OUT OF THE BLUE
SKY--
AS THE WATER VAPOR
FREEZES INTO
CRYSTALS

CUMULO-NIMULUS CLOUDS

IN THE CLOUDS
OF THE MIDDLE LEVELS
WHERE ITS WARMER
AND THERE'S MORE
WATER VAPOR..
AT A QUICKER PACE·
MORE INTRICATE
CRYSTALS OF ICE
COME TO BE

RAIN MAY COME FROM
LITTLE DROPLETS OF
WATER COALESCING
OR
RAIN MAY COME FROM
THE MELTING OF
CRYSTALS OF ICE
OR SNOWFLAKES

WATER IN THE
CLOUDS
RETURNS TO EARTH
AS RAIN

WATER IN THE AIR
TOUCHES EARTH
IN THE MIST··
AND RETURNS IN
THE EARLY MORNING
·DEW·

WATER IN SPACE
FREEZES INTO
AN INFINITE VARIETY
OF SNOWFLAKE CRYSTALS

CLOUDS FORMING
IN UP-DRAFTS

THE WINDS ARISE
AS THE AIR SWIRLS
ABOUT THE MOUNTAINS
AND MOVES
ACROSS THE VALLEYS
AND THE DESERTS
AND THE WATERS
OF THE ROTATING
EARTH·

AS THE WATERS OF EARTH
ARE WARMED··

SOME WATER EVAPORATES
AND RISES AS VAPOR·

WHICH COOLS INTO
DROPLETS··

WHICH JOIN TOGETHER
INTO
THE CLOUDS OF EARTH

TO FLOAT IN THE SKIES
TO EVENTUALLY
COOL AND RETURN
TO EARTH

AS SNOW OR RAIN

FROM THE WATER VAPOR
IN THE AIR THAT HAS RISEN
FROM OCEANS AND SEAS
AND LAKES AND RIVERS·
AND ELSEWHERE·
WHERE AND WHEN
THE AIR IS COOL ENOUGH··

CONDENSATION APPEARS
ON LITTLE CHARGED
PARTICLES AND ON DUST·
·INTO 'DROPLETS' OF WATER
LESS THAN $\frac{1}{1000}$ MM ACROSS)
(AND ON CRYSTALS OF ICE)

WHEN THE DROPLETS
BECOME BIGGER··
($\frac{1}{100}$ MM ACROSS)
A VISIBLE MIST APPEARS·
AND THE CLOUDS
COME TO BE·

WITHIN THE CLOUDS
DROPLETS FORM INTO
BIGGER DROPS·
(1 TO 5 MM ACROSS)
WHICH ARE THE
BEGINNINGS OF
RAINDROPS

THE SKIES

BLUE SKIES

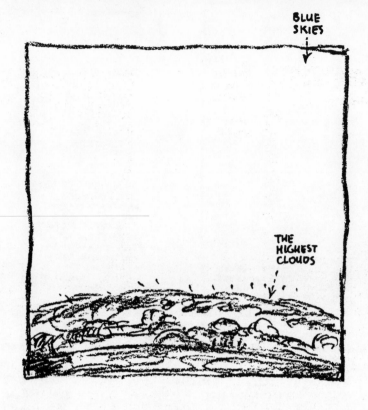

THE SKIES
NEAREST THE EARTH
ARE THE WARMEST
AND THE LIGHTEST
BLUE.

ALL THE COLORS
OF THE RAINBOW
AND THE WEATHER
ARE IN
THE SKIES
NEAREST THE EARTH

THE HIGHEST CLOUDS

THE SOLAR WIND

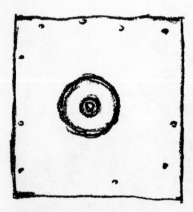

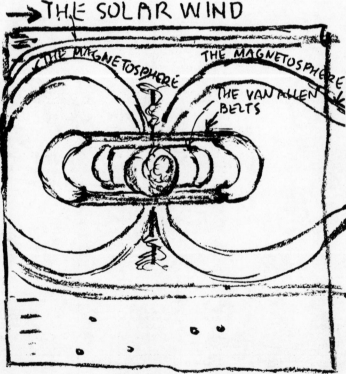

THE MAGNETOSPHERE

THE MAGNETOSPHERE

THE VAN ALLEN BELTS

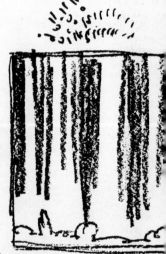

AT THE CENTER OF EARTH
IS A SOLID IRON AND NICKEL
SPHERE..
SURROUNDED BY
A MOLTEN METAL
OUTER
CENTER

THE ROTATION OF
THE PLANET EARTH
AND THE MOVING
LIQUID METAL CORE
WITHIN
PRODUCE A GREAT
MAGNETIC FIELD
AROUND THE EARTH
BETWEEN THE NORTH
AND SOUTH POLES

THE MAGNETIC FIELD
IS SHAPED BY
THE SOLAR WIND
FROM THE SUN
INTO A
TEARDROP
"MAGNETOSPHERE"

MOST OF THE
SOLAR WIND
IS DEFLECTED
AROUND THE
PLANET EARTH
BY THE
MAGNETOSPHERE

SOLAR WIND
PARTICLES
ARE HELD FOR
A WHILE IN
TWO DOUGHNUT
SHAPED
REGIONS..

"THE
VAN
ALLEN
BELTS"

SOMETIMES
SOLAR WIND PARTICLES
HIT THE VAN ALLEN BEL
IN CERTAIN WAYS
AND ARE SPEEDED UP
AND SENT TO THE
UPPER ATMOSPHERE

WHEN THEY MEET,
THERE'S A BEAUTIFUL
SHOW OF LIGHT..

THE
AURORA BOREALIS
IN THE
NORTH

THE
AURORA
AUSTRALIS
IN THE
SOUTH

THE SKIES ARE BLUE
BECAUSE THE BLUE LIGHT WAVES
OF THE LIGHT OF THE SUN
ARE SCATTERED BY THE
ATOMS OF THE ATMOSPHERE

AS THE SUN NEARS THE HORIZON
THE LIGHT PASSES THROUGH
A LOT MORE ATOMS.. AND
THE BLUE WAVES ARE SCATTERED
SO MUCH THEY BEGIN TO DISAPPEAR

WHILE THE BROADER YELLOW AND
ORANGE AND RED WAVES PASS
THROUGH RELATIVELY
UNAFFECTED

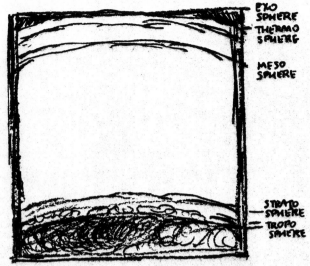

EXO SPHERE
THERMO SPHERE
MESO SPHERE
STRATO SPHERE
TROPO SPHERE

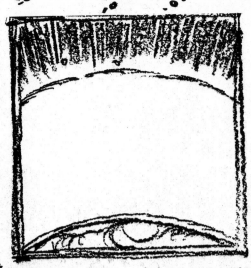

ATOMS FLOATING AWAY INTO SPACE

...AIR LAYER NEAREST ...E EARTH -- WHERE ...E WEATHER IS... ...CALLED THE ...OPOSPHERE...

...GH IN THE TROPOSPHERE, ...E SKIES ARE DEEP BLUE ...ND THE JET STREAM WINDS ...OW AT TREMENDOUS ...EEDS.

THE STRATOSPHERE IS ABOVE THE TROPOSPHERE.. FROM ABOUT 6 TO 31 MILES ABOVE EARTH.

THE STRATOSPHERE CONTAINS THE OZONE THAT SHIELDS US FROM THE ULTRAVIOLET LIGHT FROM THE SUN.

ABOVE THE STRATOSPHERE ARE THE DARK BLUE SKIES OF THE MESOPHERE..

AND THEN.. THE VERY THIN MIDNIGHT BLUE SKIES OF THE THERMOSPHERE.. --UP TO ABOUT 310 MILES ABOVE EARTH.

ABOVE THAT.. ARE THE HIGHEST REACHES OF OUR SKIES..

THE EXOSPHERE.. WHERE THE AIR IS SO THIN, AND GRAVITY IS SO WEAK THE ATOMS FLOAT AWAY INTO SPACE.

COSMIC RAYS

...ATMOSPHERE ...OMBARDED BY

..SMIC RAYS"

..MMA RAYS ..D ELECTRONS ..D NEUTRINOS ..D ALPHA ..RTICLES ..LIUM NUCLEI)

..TTING THE ..MOSPHERE ...SUCH ..REDIBLE ..ERGY THEY ..ST BE TRAVELING ...SPEEDS ..PROACHING ..T OF LIGHT

IT ALL APPEARS TO COME FROM EVERY DIRECTION, SO IT'S CALLED, "COSMIC RADIATION" OR "COSMIC RAYS"

IT'S A MYSTERY WHERE IT ALL COMES FROM

"SOME SAY COSMIC RAYS" MAY HAVE BEEN SENT FORTH BY "GALACTIC CYCLOTRONS" AT SPEEDS APPROACHING THAT OF LIGHT!

SOME SAY COSMIC RAYS COULD COME FROM QUASARS

SOME SAY LITTLE "WHITE HOLES" THROUGHOUT ALL OF SPACE ARE CONTINUALLY CREATING "LIGHT"

..F ALSO-- THERE IS THE "IONOSPHERE" BEGINNING A LITTLE ABOVE THE STRATOSPHERE AND REACHING ALL THE WAY UP TO THE EXOSPHERE

WHICH COMES FROM ELECTRICALLY CHARGED ATOMS CALLED "IONS" CREATED BY BLASTS FROM THE SUN.. BROUGHT TO EARTH BY THE SOLAR WIND.

COMETS

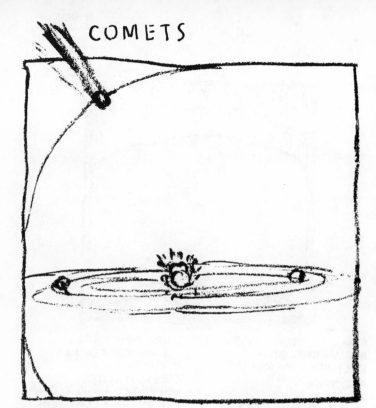

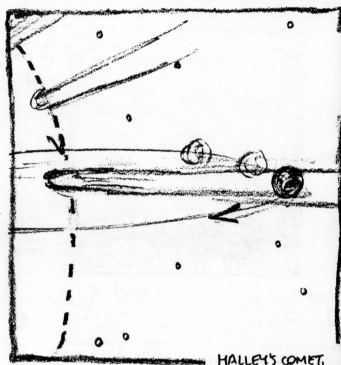

HALLEY'S COMET PASSING BETWEEN THE SUN AND THE EARTH

COMETS CIRCLE THE SUN IN HIGHLY ECCENTRIC ORBITS

SOME COMETS MAY TRAVEL IN GREAT ORBITS UP TO 12 TRILLION MILES FROM THE SUN

HALLEY'S COMET, LAST SEEN NEARBY IN MAY OF 1910, RETURNS TO THE NEIGHBORHOOD OF THE SUN EVERY 76 YEARS.

METEORS

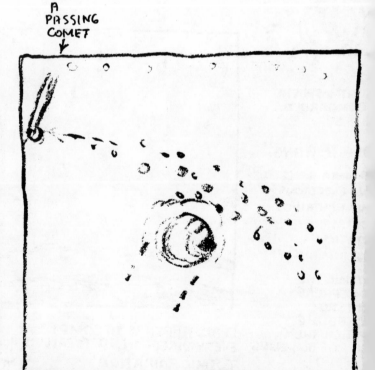

A PASSING COMET

EVERY DAY, ABOUT 200,000 METEORS "FLASH" IN THE SKIES ABOUT 70 MILES ABOVE EARTH.

POPULARLY KNOWN AS "SHOOTING STARS", METEORS ARE LITTLE ROCKS THAT VAPORIZE AS THEY HIT THE ATMOSPHERE.

CONCENTRATED METEOR SHOWERS OCCUR WHEN THE EARTH MOVES THROUGH REGIONS OF "METEOROIDS"

METEOROIDS MAY HAVE COME FROM THE WAKE OF PASSING COMETS

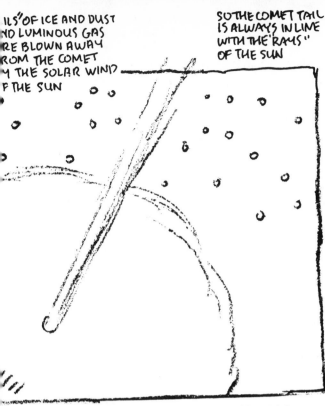

ILS" OF ICE AND DUST
ND LUMINOUS GAS
RE BLOWN AWAY
ROM THE COMET
Y THE SOLAR WIND?
F THE SUN

SO THE COMET TAIL IS ALWAYS IN LINE WITH THE "RAYS" OF THE SUN

THE "OORT" CLOUD

E NUCLEI
COMETS
E ICE AND
OZEN GASES
VERED BY
CK DUST

NEARING THE SUN, A "COMA" OF GAS AND DUST .. SOME UP TO 500,000 MILES ACROSS.. BECOMES VISIBLE ABOUT THE NUCLEUS

WITHIN ABOUT 185 MILLION MILES FROM THE SUN, LUMINOUS "TAILS" APPEAR .. SOME MORE THAN TEN MILLION MILES LONG

SOME SAY THAT COMETS COME FROM A GREAT CLOUD OF ICE AND DUST ORBITING THE SUN FAR BEYOND PLUTO .. AT THE FURTHEST REACHES OF THE SOLAR SYSTEM.

-- THAT EVERY NOW AND THEN A COMET NUCLEUS MAY BE "TAKEN" BY THE MOVEMENTS OF THE STARS

ETEORITES

LARGER ROCKS CALLED "METEORITES" COME FROM THE ASTEROIDS.

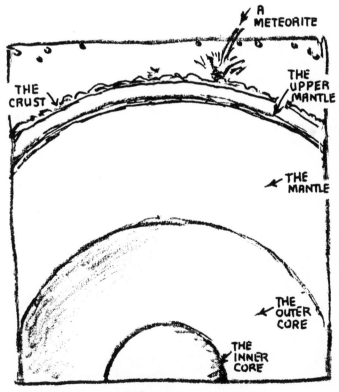

A METEORITE

THE CRUST

THE UPPER MANTLE

THE MANTLE

THE OUTER CORE

THE INNER CORE

SOME "METEORITES" ARE BIG ENOUGH TO MAKE IT THROUGH THE ATMOSPHERE AND FALL TO EARTH.

THE SUN

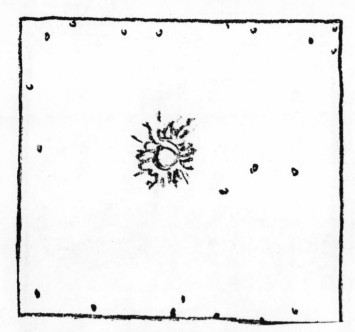

ABOUT FOUR OUT OF EVERY 100 STARS IN THE MILKY WAY GALAXY IS A STAR LIKE THE SUN

AFTER A STAR LIKE THE SUN FIRST LIGHTS IN NUCLEAR FUSION.. IT LIVES AS A BRILLIANT "BLUE STAR" FOR ABOUT 17 MILLION YEARS.

THEN IT BECOMES A MAIN SEQUENCE YELLOW STAR LIKE THE SUN.. AND LIVES AS SUCH FOR BILLIONS OF YEARS

THE SUN IS ABOUT 864,000 MILES ACROSS.. MORE THAN 100 TIMES BIGGER THAN THE EARTH

A HUGE, BILLOWING SPHERE.. MOSTLY OF HYDROGEN GAS

THE SURFACE TEMPERATURE IS ABOUT 5,700° C.

PROTONS. NEUTRONS. ELECTRONS. AND HIGH ENERGY PHOTONS FILL THE CENTER.. WHERE IT'S ABOUT 15,000,000° C.

A SOLAR ERUPTION

BRILLIANTLY FLARING SURFACES COME AND GO

 WHEN A STAR FIRST LIGHTS UP IN NUCLEAR FUSION, THE HYDROGEN "FUSIONS" INTO HELIUM... FOUR PROTONS (A HYDROGEN NUCLEUS) CHANGE INTO TWO PROTONS AND TWO NEUTRONS (A HELIUM NUCLEUS)... ...LIBERATING PHOTONS AND NEUTRINOS AND ENERGY. $E = mc^2$.

THE
SURFACE
OF
THE SUN

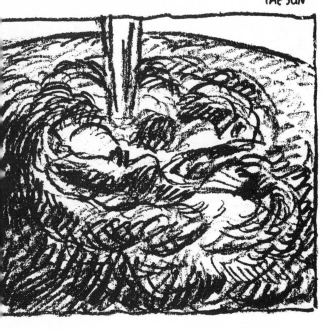

THE ENTIRE SUN
IS PULSATING LIKE
A COSMIC "BELL"..
EVERY HOUR OR TWO.

OTHER PARTS OF
THE SURFACE
PULSATE IN
VARYING PERIODS
OF TIME..
UP TO AN HOUR.

PARTS OF THE SURFACE,
LIKE WAVES ON AN OCEAN,
RISE AND FALL
ABOUT 1,000 MILES
EVERY 5 MINUTES!

WHIRLING
SOLAR TORNADOS
COME AND GO

TREMENDOUS "FOUNTAINS"
BURST FROM THE SURFACE..
UP TO 300,000 MILES
INTO SPACE!

THE
CORONA
OF
THE SUN

A GREAT SOLAR "CORONA"
RIDES THE SURFACE
OF THE SUN

SENDING FORTH
GREAT ARRAYS OF
SUB-ATOMIC PARTICLES
AND
MAGNETIC FIELDS

THAT TRAVEL
THE SOLAR SYSTEM
AS
"THE SOLAR WIND"

THE STARS

THERE ARE A GREAT VARIETY OF STARS IN THE HEAVENS...

THE STARS THAT APPEAR THE BRIGHTEST TO US ON EARTH ARE NOT THE MOST "LUMINOUS".

THE MOST LUMINOUS STARS OF THE HEAVENS ARE VERY FAR AWAY AND IN POSITIONS WHERE THEY DON'T APPEAR SO BRIGHT.

THE BRIGHTEST APPEARING STAR TO US ON EARTH IS THE STAR SIRIUS IN CANIS MAJOR -- IT APPEARS TO BE A LOT BRIGHTER THAN BRILLIANT STARS TENS OF TIMES FURTHER AWAY AND HUNDREDS OF TIMES MORE ABSOLUTELY LUMINOUS.

THE MOST LUMINOUS STARS ARE MORE THAN A MILLION TIMES MORE LUMINOUS THAN THE SUN

THE LEAST LUMINOUS STARS ARE THOUSANDS OF TIMES LESS LUMINOUS THAN THE SUN

ABOUT 75% OF THE STARS OF OUR GALAXY ARE THE LITTLE COOL STARS

"RED DWARF" STARS GIVE OFF ONLY ABOUT ONE/TEN THOUSANDTH AS MUCH LIGHT AS THE SUN...

SOME "WHITE DWARF" STARS ARE ABOUT 20,000 TIMES DIMMER THAN THE SUN...

SOME STARS ARE AS THIN AS AIR. SOME STARS THOUSANDS OF TIMES MORE DENSE THAN GRANITE.

THE DENSEST STARS ARE MORE THAN A MILLION TIMES DENSER THAN THE SUN.

SOME STARS.. VERY "THIN" SUPERGIANTS ARE ABOUT A MILLION TIMES LESS DENSE THAN THE SUN

BILLIONS OF STARS ARE SO DIM THEIR LIGHT CAN'T BE SEEN BY US.. EVEN THOUGH THEY'RE NOT TOO FAR AWAY

SOME SUPERGIANT STARS
ARE SO BIG...
BILLIONS OF STARS
THE SIZE OF THE SUN
COULD FIT
INSIDE THEM

SOME WHITE DWARF STARS
ARE SO LITTLE...
TENS OF THOUSANDS
OF THEM COULD FIT
INSIDE THE SUN

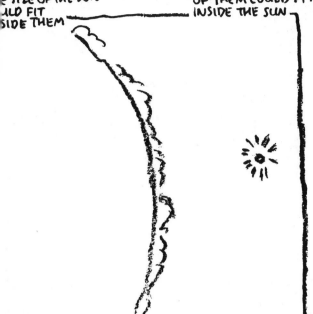
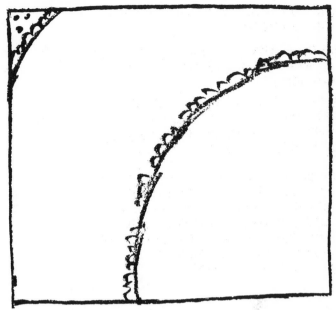

THE BIGGEST STARS
ARE ABOUT
A THOUSAND TIMES
BIGGER THAN
THE SUN

THE LITTLEST STARS
ARE ABOUT
75,000 TIMES
LITTLER THAN
THE SUN.

SOME RED GIANT STARS
WERE BORN LITTLER...
AND AFTER
BILLIONS OF YEARS
BALLOONED UP
INTO
RED GIANTS

SOME RED GIANTS
WERE ONCE
EVEN BIGGER...
AND COOLED
DOWN INTO
"RED GIANTS"

SOME "GIANT"
AND SOME
"SUPERGIANT"
STARS
PULSATE
AND VARY
IN
BRIGHTNESS
OVER TIME

STARS ARE CLASSIFIED:

O	BLUE	30,000 TO 50,000° K.
B	BLUE WHITE	11,000 TO 30,000° K.
A	WHITE	7,500 TO 11,000° K.
F	WHITE YELLOW	6,000 TO 7,500° K.
G	YELLOW	5,000 TO 6,000° K.
K	ORANGE	3,500 TO 5,000° K.
M	RED	3,500° K AND COOLER

THERE ARE
BLUE AND YELLOW
AND RED
SUPERGIANT STARS
AND
GIANT STARS

THERE ARE
BRILLIANT
WHITE
AND BLUE
STARS

AND
YELLOW AND
ORANGE AND
RED AND
BROWN STARS

THE COLOR GIVES
US AN INDICATION
OF HOW HOT THEY
ARE..
THE BLUE STARS
ARE OF THE
HIGHEST
TEMPERATURES

THE STARS OF
THE LITTLEST MASS
LIVE BILLIONS OF YEARS
LONGER THAN
THE OTHERS

A STAR WITH
10 TIMES THE MASS
OF THE SUN LIVES
FOR ABOUT
100 MILLION YEARS

A STAR LIKE
THE SUN LIVES
FOR ABOUT
10 BILLION YEARS

A STAR
$\frac{1}{10}$ THE MASS OF THE SUN
IS THOUGHT TO LIVE
FOR ABOUT
1,000 BILLION YEARS!

STARLIGHT

A PROTOSTAR LESS THAN ONE TENTH THE MASS OF THE SUN IS TOO LITTLE TO EVER LIGHT IN FUSION AS A STAR

BUT IT MAY COOL DOWN AND BECOME A PLANET

A PROTOSTAR MORE THAN 65 TIMES THE MASS OF THE SUN NEVER BECOMES A STAR

IT'S TOO BIG.. IT BECOMES UNSTABLE AND FALLS APART

WITHIN THE CLOUD IN THE BEGINNING OF A STAR

LITTLE ATOMS OF HYDROGEN MOVE CLOSER TOGETHER TO FORM SPHERES OF DENSER GAS

THAT BEGIN TO MOVE INWARD

AND GREAT PROTOSTARS COME TO BE..

THOUSANDS OF TIMES BIGGER THAN AN AVERAGE STAR

A PROTOSTAR OF A STAR THE SIZE OF THE SUN

LIVES FOR ABOUT TEN MILLION YEARS

THE SPHERE MOVES INWARD OF ITS OWN GRAVITATION

BECOMING WARMER AND WARMER AND DENSER.

THE INSIDE STORY

WHEN THE NUCLEAR FIRE IS LIT.. THE STAR MAY BE VARIABLE FOR A WHILE.. SENDING FORTH GREAT WINDS FILLED WITH STELLAR MATTER.

EVENTUALLY REACHING "THERMONUCLEAR EQUILIBRIUM".. TO LIVE FOR BILLIONS OF YEARS AS A MAIN SEQUENCE STAR

MOST KNOWN STARS ARE "MAIN SEQUENCE" STARS.. CALLED THAT WHEN THEY ARE IN THE HYDROGEN "FUSION" PROCESS WHEREIN THEY SPEND MOST OF THEIR LIVES

THE NUCLEAR "FIRE" AT THE CENTER OF MAIN SEQUENCE STARS "FUSIONS" HYDROGEN NUCLEI (PROTONS) TOGETHER LIBERATING ENERGY

IN ALL MAIN SEQUENCE CYCLE THE END RESULT IS THE SAME.. HYDROGEN INTO HELIUM RELEASING ENERGY IN ACCORDANCE WITH $E=mc^2$

DEEP WITHIN THE CLOUDS

A PROTOSTAR BEGINS TO GLOW

AS GRAVITATIONAL ENERGY CHANGES TO HEAT ENERGY

HEN THE WARD MOVEMENT BROUGHT TO REST OR A WHILE N THE OUTWARD RESSURE OF ME CENTER..

THE PROTOSTAR BEGINS TO GLOW.. AND NOW IT BECOMES VISIBLE TO US.. IN "INFRARED LIGHT" ONLY

CONTINUING ITS INWARD JOURNEY.. ILLUMINATED BY ITS OWN INWARD MOVEMENT..

THE CORE GETS WARMER AND WARMER WHILE THE SURFACE REMAINS AT ABOUT 4,000°C.

AFTER MILLIONS OF YEARS.. THE INTERIOR REACHES A GREAT ENOUGH PRESSURE AND HEAT TO "LIGHT" THE NUCLEAR FIRE AND A STAR IS BORN!

THE NEUTRINOS THAT ARE LIBERATED IN THE SAME PROCESS AT THE CENTER FLASH THROUGH EVERYTHING AT THE SPEED OF LIGHT!

T THE CENTER F A STAR LIKE HE SUN.. BOUT 5 MILLION ONS OF MATTER HANGE INTO NERGY VERY ECOND!

AS THE LIBERATED ENERGY.. OVER MILLIONS OF YEARS.. WANDERS ITS WAY OUT TO THE SURFACE IT LOSES ENERGY.. AND TRANSFORMS INTO THE ENTIRE ELECTROMAGNETIC SPECTRUM

FROM THE LIBERATED HIGH ENERGY GAMMA LIGHT INTO AND THROUGH ALL THE WAVELENGTHS OF X-RAY, ULTRAVIOLET... ALL THE COLORS OF VISIBLE LIGHT, INFRARED, MICROWAVE, AND RADIO WAVE "LIGHT"

OVER BILLIONS OF YEARS. OF FUSIONS OF FUSIONS.. THE HEAVIER ATOMS.. INCLUDING OXYGEN AND NITROGEN AND SILICON -- ALL THE ELEMENTS UP TO IRON.. ARE BROUGHT TO LIGHT FROM THE FIRE OF THE STARS.

HERE IS A PICTURE
OF HOW A STAR
LIKE THE SUN MIGHT
END ITS "DAYS"

THE SUN
FUELED
BY HELIUM
FOR A FEW
MILLION YEARS

ZOOM IN

A NEW HYDROGEN
THERMONUCLEAR
FUSION FIRE
IN A SHELL
AROUND
THE HELIUM
CENTER

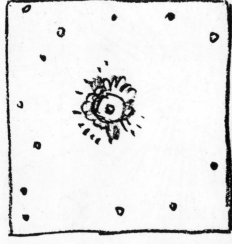

FLASH

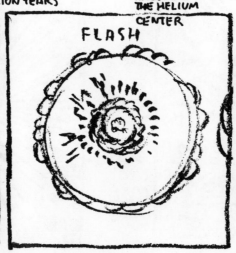

THE SUN SHOULD LIVE
PRETTY MUCH AS IT IS
FOR BILLIONS OF YEARS
UNTIL THE LAST OF
THE HYDROGEN
AT THE CENTER
IS GONE . . .

THEN, THE HELIUM
AT THE CENTER
FUELS THE STAR .

AFTER A FEW MILLION
YEARS, THE HELIUM
BECOMES SO HOT
IT IGNITES A NEW
HYDROGEN NUCLEAR
FIRE . . IN A SHELL
SURROUNDING THE
HELIUM CENTER

THE REMAINING
HELIUM FUEL
AT THE CENTER
CONTINUES ITS
INWARD MOVE . .
GETTING HOTTER
AND HOTTER . .

WHILE THE ENERGY
FROM THE
HYDROGEN
BEGINS ITS TRIP
TO THE SURFACE

AFTER MANY YEARS.
THE ENERGY FROM
THE NEWLY LIT
HYDROGEN SHELL
ARRIVES AT
THE OUTER REGIONS

PUSHING THE
ATMOSPHERE
MILLIONS OF MILES
INTO SPACE !

THE OUTER REGIONS
BALLOONING OUTWARD
TO EVENTUALLY REACH
A SIZE HUNDREDS OF
TIMES THAT OF
THE SUN . .
TO BECOME A
"RED GIANT"

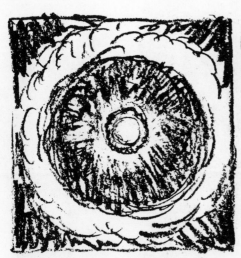

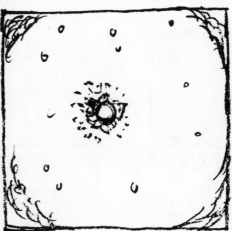

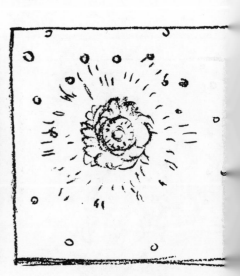

AFTER A WHILE
THE STAR
BEGINS ITS
FINAL
INWARD
JOURNEY

AS THE
OUTER LAYERS
FLOAT OFF
INTO THE
VASTNESS
OF
SPACE

NOW
FUELED
BY
HELIUM
BURNING"
ABOUT A CENTER
OF CARBON
AND OXYGEN

THE STAR
MOVES
INWARD

AS TIME PASSES.
THE COOLING STAR
·PULSATES·
· · · ·
IN BRIGHTNESS
AND
SIZE *

* ' AS A "CEPHEID VARIABLE"
AND/OR AN "RR LYRAE"
VARIABLE STAR ?

THE OUTER REGIONS
ZOOMING INWARD
IGNITE A
HELIUM
THERMONUCLEAR
FUSION FIRE "

FLASH

ZOOM

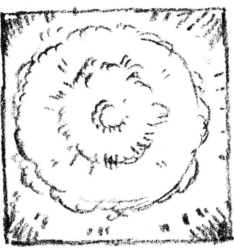

TER PERHAPS
BILLION YEARS,
HE HYDROGEN OF
E CENTER SHELL
USED UP.

NE NUCLEAR FIRE
OES OUT...
ND THE OUTER
EGIONS
OOM INWARD...

PUSHING THE
HELIUM CENTER
FURTHER INWARD..
TILL IT GETS UP
TO 200 MILLION
DEGREES..
AND THE HELIUM
IGNITES IN
NUCLEAR FUSION..
CALLED
THE HELIUM FLASH *2

THIS HALTS THE
INWARD MOVE
OF THE HELIUM
AND SENDS
FORTH
TREMENDOUS
ENERGY
ACCORDING TO
$E = mc^2$

THE OUTER REGIONS
ZOOM OUTWARD AGAIN.

THIS TIME
A HUNDRED TIMES
FASTER THAN BEFORE..
INTO A SECOND
"RED GIANT"
PHASE

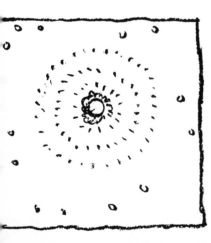

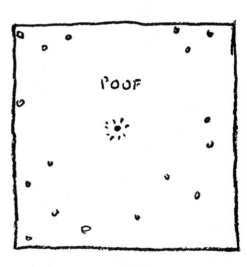

POOF

THE STAR
THEN COMES
TO LIVE
FOR MILLIONS OF
YEARS AS A
"WHITE DWARF"
STAR.

A "WHITE DWARF"
STAR IS A VERY
DENSE, BRIGHT
STAR..
ABOUT AS BIG
AS
PLANET EARTH

A STAR
WHEREW THE
NUCLEI ARE
PUSHED TOGETHER
TO THE
LIMITS OF STRENGTH
OF THE MINIMUM
ATOMIC ORBITS

AS TIME PASSES,
THE STAR DIMS..
AND OVER THE
YEARS..
IT BECOMES
YELLOW...
THEN
RED..

AFTER
MILLIONS OF YEARS.
OF THE LAST
GLOWINGS..

ITS LUMINOUS DAYS
COME TO AN
END.

---

2 IN THE
"HELIUM
FLASH"

3 HELIUM
NUCLEI
FUSE
INTO

ONE
CARBON
NUCLEUS
(6 PROTONS
AND 6 NEUTRONS)

LIBERATING
TREMENDOUS
ENERGY

FLASHING STARS

THERE ARE A GREAT MANY KINDS OF PULSATING STARS AND FLASHING STARS . . .

MOST PULSATING VARIABLE STARS BRIGHTEN AND DIMINISH AT REGULAR INTERVALS.

THE "RR LYRAE VARIABLE" STARS CHANGE IN BRIGHTNESS IN PERIODS OF LESS THAN A DAY.

THE "CEPHEID VARIABLE" STARS CHANGE IN BRIGHTNESS IN PERIODS OF FROM 3 DAYS TO ABOUT 50 DAYS.

LONG PERIOD PULSATING VARIABLES, WHICH ARE USUALLY GIANT STARS IN THEIR LATER YEARS, PULSATE IN PERIODS OF UP TO 1,000 DAYS

SOME GIANT "CEPHEID VARIABLE" STARS ARE HUNDREDS TO THOUSANDS OF TIMES BRIGHTER THAN THE SUN

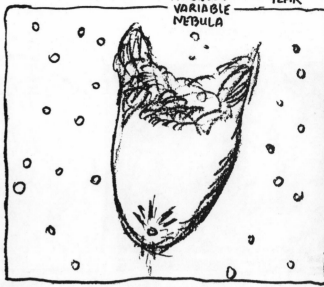

A VARIABLE STAR AT THE TIP OF HUBBLE'S VARIABLE NEBULA

VARIES IN BRIGHTNESS YEAR TO YEAR

WE CAN SEE FLASHINGS, AND DIMMINGS AND BRIGHTEN

BECAUSE THE STAR ITSELF IS INTRINSICALLY FLASHING OR INTRINSICALLY PULSATING

BECAUSE THERE'S TREMENDOUS ENERGY EXCHANGES BETWEEN STARS

BECAUSE DOUBLE AND OR MULTIPLE STAR SYSTEMS ARE ECLIPSING

BECAU' THERE . CLOUDS MOVIN' IN FRO' A STAR OR STA'

FLASHING DOUBLE STARS

SOME "LITTLE CLOSE BINARIES" ECLIPSE IN DAYS

SOME SUPERGIANT "CLOSE BINARIES" ECLIPSE IN YEARS

"SOME VERY CLOSE BINARIES" HAVE EGG-SHAPED STARS

SOME "CONTACT BINARIES" TOUCH ONE ANOTHER.

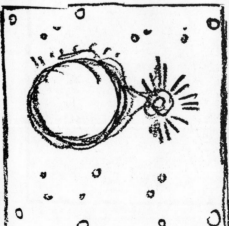

THERE ARE A GREAT MANY DOUBLE STAR SYSTEMS

SOMETIMES DOUBLE STAR SYSTEMS.. CALLED "BINARY STARS" ARE BORN TOGETHER.. SOMETIMES THEY CAPTURE EACH OTHER IN CLOSE ENCOUNTERS AND BEGIN TO ROTATE ABOUT EACH OTHER

ECLIPSING BINARY STARS FLUCTUATE IN BRIGHTNESS . .

APPEARING FROM AFAR TO BE LIKE THE INTRINSICALLY VARIABLE STARS

SOME ASTRONOMERS THINK THAT THE EVOLUTIONARY PHASES OF CERTAIN BINARY SYSTEMS EXPLAIN MANY MYSTERIOUS IRREGULAR VARIABLE STAR RADIATIONS

A HIGH LUMINOSITY "X RAY VARIABLE STAR" COULD BE A BINARY SYSTEM.. A NEUTRON STAR AND A MASSIVE STAR FEEDING ENERGY BACK AND FORTH.. ROTATING ABOUT EACH OTHER..

A "GAMMA RA' VARIABLE STA' COULD BE A BINARY SYSTEM

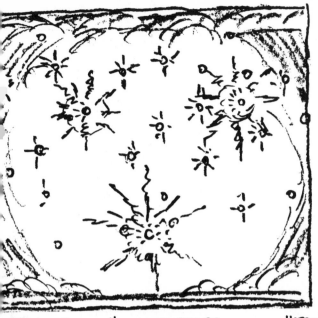 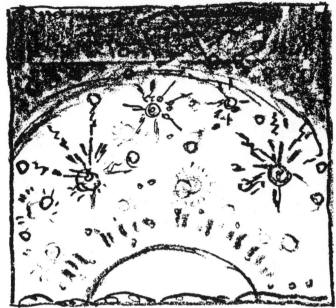

...ME STARS "FLASH"
...AMAZINGLY
...GULAR CYCLES ..

...ME STARS "FLASH"
...IRREGULAR
...AYS .

SOME STARS . . . .
FLARE BRILLIANTLY
EVERY NOW AND
THEN .

"ERUPTIVE VARIABLES"
FLARE FORTH, IN
TEMPORARY "BURSTS"

SOME "X-RAY BURSTERS"
FLARE FOR A
FEW SECONDS
AT REGULAR INTERVALS
HOURS APART

ABOUT A HUNDRED
VERY BRIGHT
VARIABLE
X-RAY STARS HAVE
BEEN DISCOVERED
IN OUR OWN GALAXY...
..SEVEN OF THEM ARE
WITHIN GLOBULAR
CLUSTERS IN THE
GALACTIC HALO

THESE STARS
RADIATE
THOUSANDS OF
TIMES
MORE POWER
IN THE X-RAY
WAVELENGTHS
THAN THE SUN
RADIATES IN
ALL
WAVELENGTHS

NOVAS

 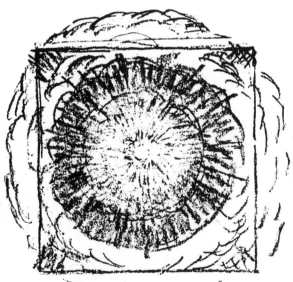

...OVAS "
...E STARS THAT
...ASH FORTH
...CREASING
...HEIR BRIGHTNESS
...O TO THOUSANDS
...TIMES BRIGHTER ..

...ND THEN RETURN
...THE WAY THEY
...ERE .

NOVAS
OCCUR
MUCH MUCH
MORE OFTEN
THAN
SUPERNOVAS

SOME SAY MOST NOVAS
COME FROM DRAMATIC
INTERCHANGES BETWEEN
BINARY STARS

WHEREIN, OVER MANY YEARS,
A LITTLE "WHITE DWARF"
COMPANION STAR
ACCUMULATES A COVER
OF GAS FROM THE
BIGGER STAR ..

OTHERS SAY
NOVAS
COME FROM
SINGLE STARS
IN THEIR
LATER
YEARS

EVENTUALLY BUILDING UP
ENOUGH PRESSURE TO
LIGHT, THE OUTER REGIONS
IN: FUSION -

EXPLODING STARS

THE MOST MASSIVE STARS
SHINE A RELATIVELY SHORT TIME
AND END IN THE MOST DRAMATIC WAYS.

A "LITTLE" SUPERGIANT STAR
ENDS ITS LIFE IN A "SUPERNOVA"...
LEAVING NOTHING BEHIND BUT DUST.

A MEDIUM SUPERGIANT STAR
BEGINS ITS END IN A SUPERNOVA..
AND BECOMES A "NEUTRON STAR"...
A "PULSAR".

A LARGE SUPERGIANT STAR
BEGINS ITS END IN A SUPERNOVA..
AND BECOMES A "BLACKHOLE"

ASTRONOMERS HAVE RECORDED
FOUR KNOWN SUPERNOVAS
IN THIS GALAXY
IN OUR TIME

SUPERNOVAS
RELEASE HUNDREDS
TO MILLIONS TIMES
MORE ENERGY
THAN NOVAS

SOME SUPERNOVAS
APPEAR TO BE
THOUSANDS OF TIMES
BRIGHTER THAN
AN ENTIRE GALAXY
OF BILLIONS OF STARS.

SOMETIMES
IF THE STAR
IS BIG ENOUGH..
THERE IS
A STAR MASS
THAT SURVIVES
THE SUPERNOVA

PULSAR

A NEUTRON STAR..
OR "PULSAR"..
COULD COME FROM
A SURVIVING MASS
OF FROM 1.4 TO 3.2
TIMES THE MASS OF
THE SUN

A PULSAR
IN THE
CRAB NEBULA

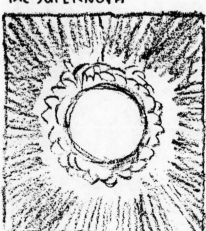

A "MEDIUM"
SUPERGIANT
WOULD GO THROUGH
THE SAME PROCESS -
BUT THERE WOULD
BE ENOUGH MASS TO
SURVIVE THE
EXPLOSION ...

THE SURVIVING MASS
COLLAPSES INTO A
RAPIDLY SPINNING
"NEUTRON STAR."
A STAR OF SOLID
NEUTRONS..
PUSHED TO THE
LIMIT OF STRENGTH
OF THE INTERNAL
FORCE OF THE
NEUTRON.

A NEUTRON STAR,
OR "PULSAR"..
IS ONLY ABOUT
TEN MILES ACROSS..
SO DENSE THAT
A HANDFUL OF
THE STAR WOULD
WEIGH MORE
THAN A
MOUNTAIN !

MOST "PULSARS"
FLASH AT REGULAR
INTERVALS..
THE PULSAR AT
THE CENTER OF
THE CRAB NEBULA
FLASHES AT A
RATE OF
30 TIMES
A SECOND

IT SENDS FORTH
GAMMA RAY LASER LIG
RADIO LIGHT, X RAY LI
AND MORE..

WITH A MAGNETIC FIEL
ONE TRILLION TIMES TH
OF EARTH .. IT CREATE
"RELATIVISTIC" ELECTRON
TRAVELING NEAR
THE SPEED OF LIGHT

SUPERNOVA

"A "LITTLE" SUPERGIANT STAR

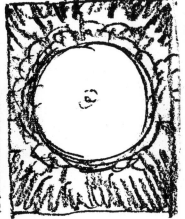

THE SUPERGIANT SOARING OUTWARD

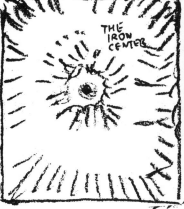

TURNING BACK IN

THE IRON CENTER

"LITTLE" SUPERGIANT A PROTOSTAR FOR ONLY 100,000 YEARS. AND LIVES AS A STAR OR ONLY A MILLION YEARS .. SHINING IN A BRILLIANT BLUE WHITE LIGHT 100,000 TIMES BRIGHTER THAN THE SUN.

WHEN THE STAR HAS USED UP ALL ITS HYDROGEN IT BEGINS TO COOL AND EXPAND OUTWARD .. NOW BURNING ITS HELIUM AND HEAVIER ELEMENTS

THE STAR BECOMES SO COOL THAT THE CENTER BECOMES IRON AND THE FIRE GOES OUT AT THE CENTER

WHEREUPON THE OUTER REGIONS TURN BACK TO ZOOM INWARD

PLUMMETING INTO A TREMENDOUS EXPLOSION .. A SUPERNOVA! LEAVING BEHIND NOTHING BUT GAS AND DUST.

A BLACKHOLE COULD COME FROM A SURVIVING MASS OF MORE THAN 3.2 TIMES THE MASS OF THE SUN.

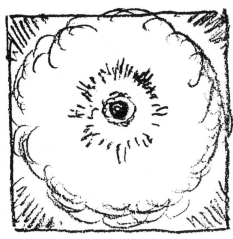

THE GREATEST OF THE SUPERGIANTS WOULD LIGHT FOR RELATIVELY FEW YEARS .. AND THEN GO TO A ... SUPERNOVA EXPLOSION

IF THE SURVIVING MASS IS GREAT ENOUGH .. IT WOULD ZOOM INWARD WITH SUCH GRAVITATIONAL STRENGTH

IT WOULD CRUSH ALL MATTER BEYOND THE RESISTANCE OF THE "STRONG FORCE" OF THE NEUTRON .. OUT OF SPACE-TIME AS WE KNOW IT.

SOME SAY IT BECOMES AN ASTRONOMICAL "BLACK HOLE" .. A "SINGULARITY" * SURROUNDED BY AN EVENT HORIZON ..
* BEYOND DESCRIPTION BY THE KNOWN LAWS OF SCIENCE

ASTRONOMICAL BLACKHOLES, IF THEY EXIST. WOULD BE INVISIBLE TO US. ASTRONOMERS MAY HAVE FOUND ONE IN CYGNUS .. BY OBSERVING WHATS GOING ON AROUND IT.

THE GALAXIES

GALAXIES

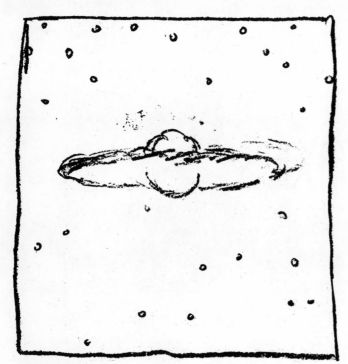

A GALAXY IS
A COMMUNITY OF
HUNDREDS OF
THOUSANDS
OF STARS

IN ALL THE UNIVERSE
THERE MAY BE
BILLIONS
OF GALAXIES

BRILLIANT CLUSTERS
TRAVEL IN HIGHLY
INCLINED AND
ECCENTRIC ORBITS

WITHIN A GREAT
"HALO" OF STARS

THE STARS NEAR THE SUN
MOVE ABOUT THE CENTER
EVERY 225 MILLION YEARS
OR SO

THE FURTHER OUT THE STAR
ARE, THE MORE SLOWLY
THEY MOVE ABOUT THE
CENTER

THE GALAXY
HAS A CENTRAL
DOME-LIKE
"BULGE"
MOSTLY OF
OLD STARS

AROUND THE
CENTRAL DOME
IS A GREAT
"HALO"
FILLED WITH
ANCIENT LITTLE
"WHITE" STARS

WITHIN THE
HALO,
THE GREAT
GLOBULAR
CLUSTERS
LIGHT THE SKIES
LIKE BRILLIANT
SHIMMERING
CRYSTAL
SPHERES

IN A
HALO
OF
STARS

THE CENTRAL
DOME IS
SURROUNDED
BY A ROTATING
"DISC" OF
BILLIONS OF
STARS
OF
A GREAT VARIETY
OF AGES AND
COLORS

THE ROTATING
DISC SPREADS
OUT INTO THE
RICH CLOUDS
OF THE ARMS
OF THE SPIRAL

-- BRILLIANTLY
ILLUMINATE
BRIGHT NEV
BABY BLUE
STARS

THE STARS OF THE GALAXY
SEEN FROM WITHIN ... THE MILKY WAY

WE LIVE IN A GALAXY
CALLED
"THE MILKY WAY"
"GALAXY"

THE
MILKY WAY
GALAXY..
ABOUT
100,000
LIGHT YEARS
ACROSS
AS
IT MIGHT
APPEAR
EDGE-ON
FROM OUTER
SPACE

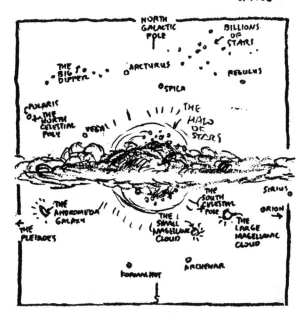

ITS THOUGHT
TO BE
A SPIRAL
GALAXY
OF STARS
OF
ALL AGES.

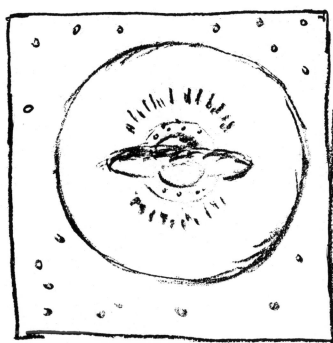

SURROUNDING THE GALAXY
IS A GREAT HALO..
200,000 LIGHT YEARS
ACROSS

THIS GALAXY OF GALAXIES "THE LOCAL GROUP"

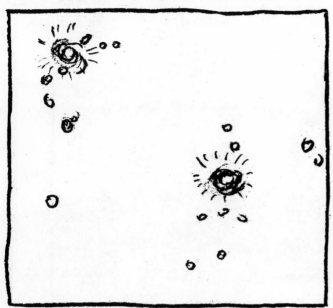

THE ANDROMEDA GALAXY IS ABOUT TWICE AS BIG AS THE MILKY WAY GALAXY. THEY ARE THOUGHT TO LOOK VERY MUCH LIKE ONE ANOTHER

THE MILKY WAY GALAXY IS ONE OF "THE LOCAL GROUP" OF TWENTY (?) GALAXIES OR MORE

MOVING TOGETHER IN A REGION OF THE HEAVENS ABOUT 2,500,000 LIGHT YEARS ACROSS

THE MILKY WAY AND THE ANDROMEDA GALAXY ARE THE TWO GREAT SPIRAL GALAXIES OF "THE LOCAL GROUP"

THEY ROTATE IN COMPLEMENTARY DIRECTIONS.. LIKE BRILLIANT COSMIC PINWHEELS

EACH IS SURROUNDED ITS OWN LITTLE GROUP OF GALAXIES..

TWO GOOD SIZED COMPANIONS AND SOME LITTLER ONES

ITS DIFFICULT TO KNOW EXACTLY HOW MANY GALAXIES ARE IN THE LOCAL GROUP, AS VERY LITTLE GALAXIES WOULD NOT BE BRIGHT ENOUGH TO BE VISIBLE IN THE FURTHER REGIONS

SOME OF THE GALAXIES IN "THE LOCAL GROUP"

THE MILKY WAY GALAXY

THE LARGE MAGELLANIC CLOUD GALAXY

THE SMALL MAGELLANIC CLOUD GALAXY

THE URSA MINOR SYSTEM

THE SCULPTOR SYSTEM

THE DRACO SYSTEM

THE FURNAX SYSTEM

THE LEO II SYSTEM

THE LEO I SYSTEM

THE GALAXY NGC 6822

THE GALAXY NGC 147

THE GALAXY NGC 185

THE GALAXY NGC 205

THE GALAXY NGC 221, M32

THE GALAXY IC 1613

THE ANDROMEDA GALAXY

THE GALAXY NGC 598, M33

THE MAFFEI 1 GALAXY

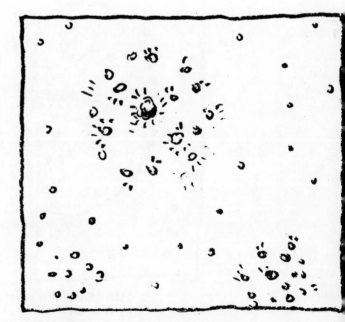

THOUSANDS OF LITTLE "GROUPS" OF GALAXIES CAN BE SEEN IN THE TELESCOPES OF TODAY.

THE MILKY WAY WITH ITS OWN LITTLE GROUP OF GALAXIES

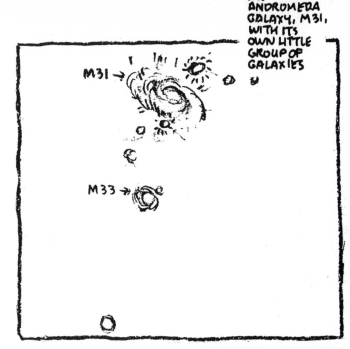

THE ANDROMEDA GALAXY, M31, WITH ITS OWN LITTLE GROUP OF GALAXIES

M31 →

M33 →

WO GALAXIES EST TO THE Y WAY..

ABOUT OOO T YEARS Y..ARE

THE "LARGE MAGELLANIC CLOUD".. AN IRREGULAR GALAXY ABOUT 40,000 LIGHT YEARS ACROSS AND THE "SMALL MAGELLANIC CLOUD" .. AN IRREGULAR GALAXY ABOUT 20,000 LIGHT YEARS ACROSS

THE ANDROMEDA GALAXY, M31, IS ABOUT 2 MILLION LIGHT YEARS AWAY

ITS COMPANION GALAXIES ARE THE GALAXIES M32 AND NGC 205, BOTH LITTLE ELLIPTICAL GALAXIES

THERE'S A 3RD SPIRAL IN THE LOCAL GROUP.. NOT AS BIG AS THE OTHER TWO.. - THE GALAXY M33 IN TRIANGULUM ABOUT 2.3 MILLION LIGHT YEARS AWAY

THE REST OF THE LOCAL GROUP ARE LITTLE ELLIPTICAL GALAXIES AND IRREGULAR GALAXIES

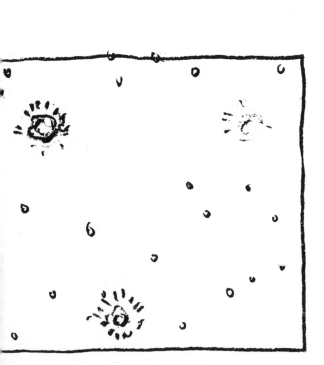

A GALAXY OF GALAXIES OF GALAXIES

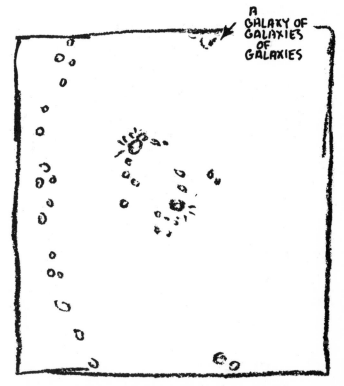

ALAXIES ARE, ON THE AVERAGE, BOUT THREE MILLION LIGHT YEARS PART

THE "LOCAL CLUSTER" IS THOUGHT TO BE NEAR THE EDGE OF A LOCAL "SUPERCLUSTER" OF GALAXIES... MORE THAN A HUNDRED MILLION LIGHT YEARS ACROSS!

SPIRAL GALAXIES

SOME GALAXIES ARE SPIRALS..
A GREAT COMMUNITY OF STARS
FROM THE VERY YOUNGEST
TO THE OLDEST

YOUNG BLUE STARS ARE BORN
IN THE RICH CLOUDLIKE
SPIRAL "ARMS"

SPIRALS CAN LOOK QUITE DIFFERENT TO US
DEPENDING UPON THE VIEW WE
SEE THEM FROM
AND THE TYPE OF SPIRAL THEY ARE

THE SOMBRERO GALAXY IN VIRGO
M104, NGC 4594   TYPE Sa/Sb

THE SPIRAL GALAXY
NGC 6946 IN CYGNU
TYPE Sc

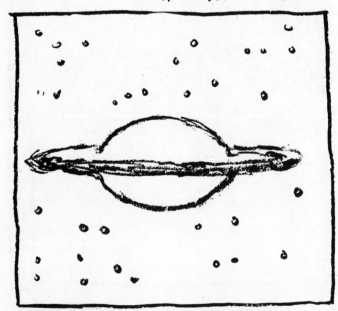

SPIRAL GALAXIES
SEEN "EDGE-ON"
LOOK QUITE
DIFFERENT THAN
WHEN SEEN FROM
ABOVE

THE "SOMBRERO"
GALAXY, M104,
IN VIRGO,
IS A SPIRAL THAT
WE SEE "EDGE-ON"

WE CAN SEE
MANY SPIRALS
FROM ABOVE

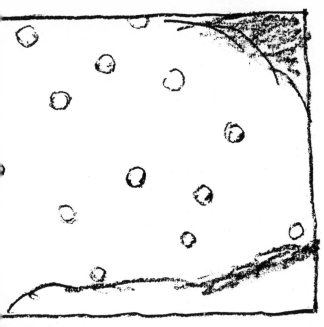

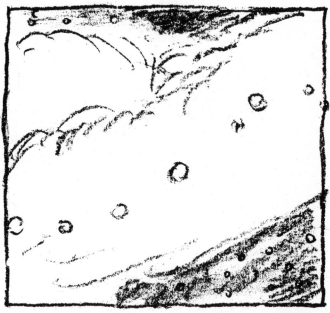

...IG OLD STARS
...ILL THE SKIES
...F THE CENTRAL
...OMES

WITHIN SPIRAL GALAXIES
MOST STARS ORBIT
THE CENTER IN
NEARLY CIRCULAR
PATHS

BOBBING UP AND DOWN
IN THE "DISC" AND IN
THE TRAILING
SPIRAL ARMS

THE SPIRAL GALAXY
NGC 7331 IN PEGASUS
TYPE Sb

THE BARRED SPIRAL
GALAXY NGC 1530
IN CAMELOPARDIS
TYPE SBb

...ND MANY FROM
...IEWS IN BETWEEN
...DGE ON AND
...BOVE

SPIRALS ARE
CLASSIFIED
INTO "TYPES"
ACCORDING
TO HOW
"TIGHTLY WOUND"
THEY APPEAR

SOME HAVE
BROAD AND
FAR REACHING
SPIRAL ARMS

SOME GALAXIES ARE
"BARRED" SPIRALS..
WHEREIN THE ARMS
APPEAR TO FLOW
FROM A "BAR"
THROUGH THE
CENTER ..

THE REASON
THE "BAR" IS
THERE
IS QUITE A
MYSTERY

ELLIPTICAL GALAXIES

A "SUPERGIANT" ELLIPTICAL GALAXY IS ABOUT 300,000 LIGHT YEARS ACROSS AND LIT BY A THOUSAND TIMES MORE STARS THAN THE AVERAGE GALAXY

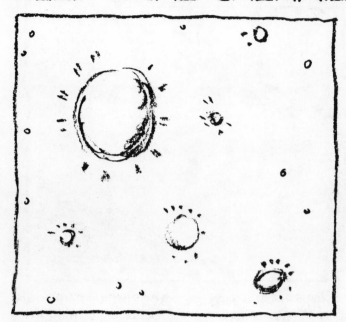

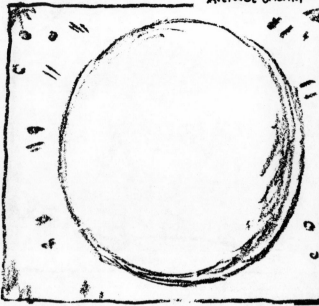

"ELLIPTICAL" GALAXIES CONTAIN MOSTLY OLD STARS.. AND NOT MUCH GAS AND DUST

SOME ARE ALMOST SPHERICAL.. LIKE THE GALAXY M49, NGC 4472 IN VIRGO

SOME ARE QUITE ELLIPTICALLY SHAPED.. SOME APPEAR TO BE A LITTLE "FLATTENED OUT"

THERE ARE MORE ELLIPTICAL GALAXIES THAN SPIRALS.. AND THE ELLIPTICALS HAVE A GREATER RANGE OF BRIGHTNESS AND SIZE

THERE ARE "GIANT" AND "SUPERGIANT" ELLIPTICALS

SOME "GIANT" ELLIPTICALS ARE BRIGHTER THAN ANY KNOWN SPIRAL GALAXY

IRREGULAR GALAXIES

A GALAXY WITH DUST LANES THAT ARE NOT THE NORMAL SPIRALS.. THE IRREGULAR GALAXY NGC 3077, TYPE II, IN URSA MAJOR

THE LARGE MAGELLANIC CLOUD IN DORADO A TYPE SBm GALAXY.. CAN BE SEEN WITH THE UNAIDED EYE FROM THE SOUTHERN SKY

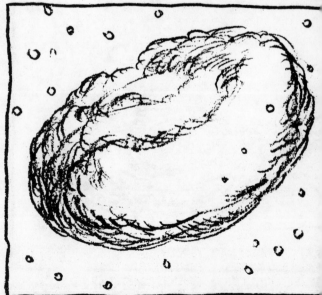

SOME GALAXIES ARE "IRREGULAR" GALAXIES

"IRREGULAR" GALAXIES ARE FILLED WITH YOUNG STARS

OUR NEIGHBOR GALAXY "THE LARGE MAGELLANIC CLOUD" IN DORADO IS AN IRREGULAR GALAXY

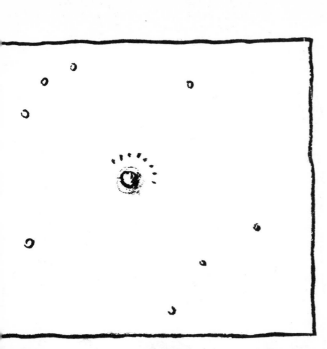

A BRIGHT BLUE JET .. OUT TO ABOUT 4,000 LIGHT YEARS FROM THE CENTER

ANY WARF" ELLIPTICALS RE SO DIM AND THIN HEY'RE BARELY ISIBLE

THERE MAY BE A LOT OF LITTLE ELLIPTICALS AND OTHER GALAXIES THROUGHOUT THE SKIES

THAT CAN'T BE SEEN WITH THE TELESCOPES OF TODAY

A GIANT ELLIPTICAL GALAXY IN VIRGO IS SURROUNDED BY A HALO OF MORE THAN A THOUSAND GLOBULAR STAR CLUSTERS

AT THE CENTER OF THE GALAXY IS A TREMENDOUS RADIO AND X-RAY SOURCE

THE GALAXY SEEMS TO BE THE SOURCE OF A GREAT MYSTERIOUS "JET"

EYFERT GALAXIES

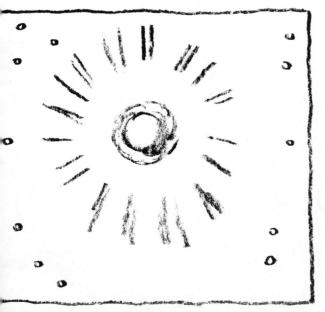

A VIOLENT SEYFERT NUCLEUS WITHIN A DISTORTED RING GALAXY .. THE PECULIAR RING GALAXY VV285

OUT 2 PERCENT ALL GALAXIES RE "SEYFERT" ALAXIES...

TYPE OF "SPIRAL" TH A VERY VERY IGHT NUCLEUS ND VERY DIM INT LOOKING ARMS

A "SEYFERT" GALAXY GIVES OFF ABOUT 100 TIMES THE RADIATION OF OUR OWN GALAXY ..

MOSTLY IN INFRARED LIGHT .. SOME ALSO EMIT STRONG X-RAY AND RADIO LIGHT

THE CENTER OF A SEYFERT GALAXY .. ABOUT 10 LIGHT YEARS ACROSS... IS LITTLE COMPARED WITH OUR GALAXY'S CENTER

CLOUDS OF GAS SOAR OUTWARD FROM A SEYFERT CENTER AT WHAT APPEARS TO BE SPEEDS UP TO THOUSANDS OF MILES A SECOND

SOME "SEYFERT" NUCLEI HAVE BEEN FOUND IN OTHER TYPES OF GALAXIES

UNUSUAL GALAXIES

THE PECULIAR GALAXY
NGC 2545

THE PECULIAR GALAXIES
NGC 3986/3988

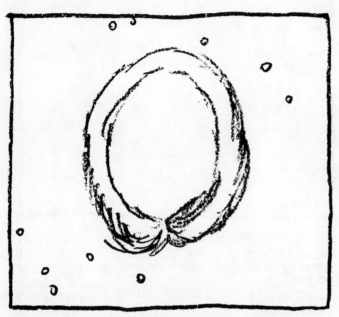

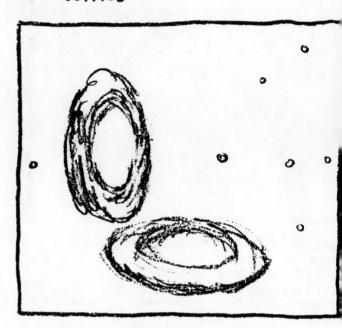

MANY GALAXIES ARE CALLED "PECULIAR"
AS NOT MUCH IS KNOWN ABOUT HOW
THEY MIGHT HAVE FORMED . .
OR . . DO SOME LOOK THE WAY THEY DO
BECAUSE WE DON'T HAVE A CLEAR VIEW ?

THE PECULIAR GALAXY
NGC 2685 IN URSA MAJOR
SEEMS TO HAVE TWO
"AXES OF SYMMETRY"

IT LOOKS SOMETHING LIKE
A SPIRAL GALAXY
SURROUNDED BY
CLOUD-LIKE STREAMERS

COLLIDING GALAXIES
NICKNAMED "THE MICE"

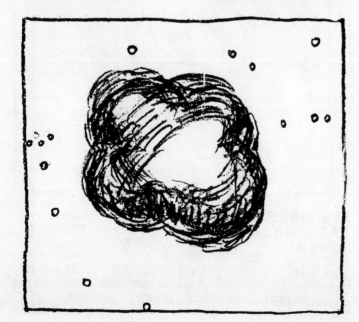

PECULIAR GALAXIES
GC 6621/6622

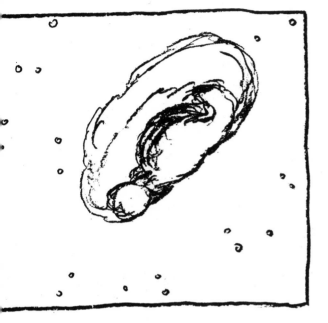

A RING GALAXY
(SOME THINK THIS RESULTS
FROM THE COLLISION OF
TWO GALAXIES)

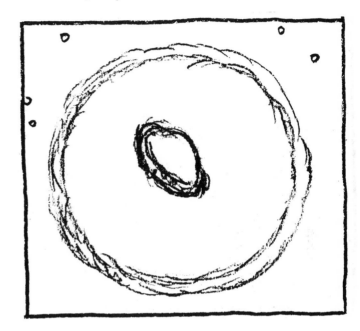

THE PECULIAR GALAXY
NGC 3718 IN URSA MAJOR
TYPE S0 (PECULIAR)

THE PECULIAR GALAXY NGC 2146
IN CAMELOPARDALIS
TYPE S ab (PECULIAR)

EXPLODING GALAXIES

SOME GALAXIES APPEAR TO BE EXPLODING

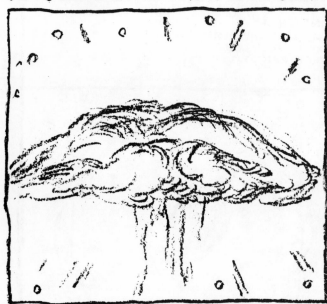

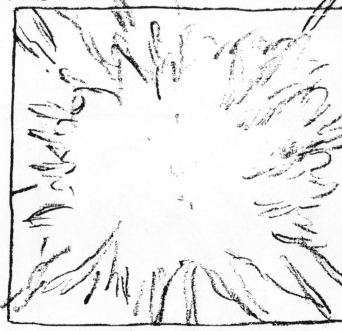

AN EXPLODING GALAXY CAN BE AS BRIGHT AT THE CENTER AS A HUNDRED MILLION SUPERNOVAE

THE EXPLODING GALAXY NGC 1275, PERSEUS A, VIEWED IN RED LIGHT, LOOKS SOMETHING LIKE THE CRAB NEBULA, BUT 10000 TIMES BIGGER

THE NUCLEUS RADIATES INTENSELY IN RADIO, INFRARED, AND X-RAY LIGHT. GAS FILAMENTS FILL THE SKY OUTWARD TO TENS OF THOUSANDS OF LIGHT YEARS AWAY

INVISIBLE GALAXIES

SOME GALAXIES CANT BE SEEN IN VISIBLE LIGHT, BUT CAN BE "SEEN" IN OTHER WAVELENGTHS

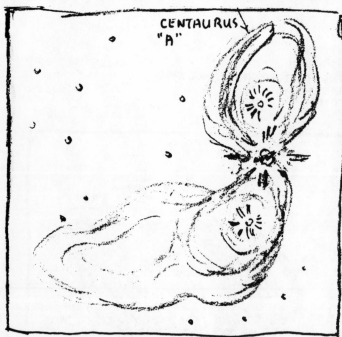

CENTAURUS "A"

RADIOLIGHT FROM THE GIANT ELLIPTICAL GALAXY IC 4296

ABOUT 120 MILLION LIGHT YE AWAY IN THE CENTAUR CLUSTER GALAXIE

TWO JETS FROM THE CENTER OF THE GALAXY (ABOUT ONE MILLION LIGHT YEARS ACROSS)

RADIOLIGHT JETS FROM ELLIPTICAL GALAXY NGC 1265

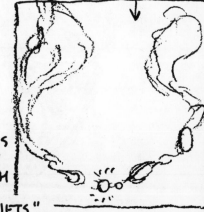

A TINY PART OF THE GALAXY "CENTAURUS A" APPEARS IN THE VISIBLE LIGHT WAVELENGTHS

A RADIOLIGHT VIEW SHOWS THAT THE OPTICAL PART LIES WITHIN A GREAT DOUBLE RADIOLIGHT SOURCE, A HUNDRED TIMES BIGGER

THE STRONGEST AREAS OF THE RADIO SOURCES ARE CENTERED UPON TWO GREAT RADIO EMITTING CLOUDS..ABOUT ONE MILLION LIGHT YEARS APART

SOME GALAXIES APPEAR TO BE SENDING FORTH GREAT RADIOLIGHT "JETS"

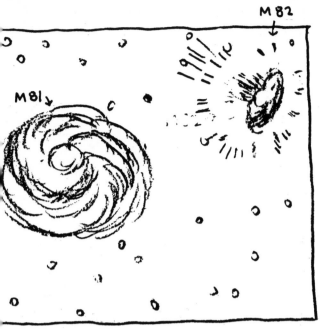

THE GALAXY M82 IN URSA MAJOR
APPEARS TO BE "EXPLODING"...
GIVING OFF GREAT AMOUNTS OF ENERGY

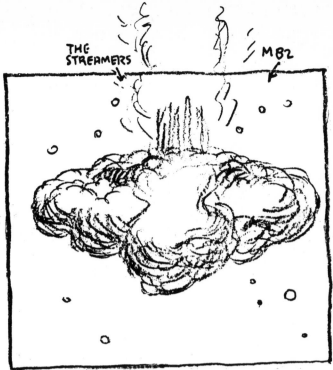

GREAT AMOUNTS
OF LIGHT COME FROM
HIGH ENERGY
ELECTRONS
SPIRALING OUTWARD
UPON THE GALAXY'S
ELECTROMAGNETIC
"STREAMERS"

AT THE CENTER, THERE IS
A STRONG RADIO WAVE SOURCE
ABOUT ONE LIGHT YEAR ACROSS

THAT SENDS "BLASTS"
ABOUT ONCE EVERY TEN YEARS
OUTWARD INTO A GREAT HALO

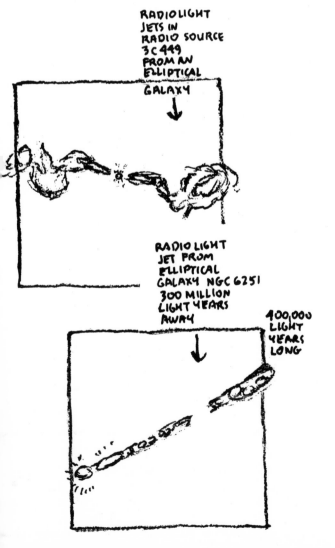

RADIOLIGHT
JETS IN
RADIO SOURCE
3C 449
FROM AN
ELLIPTICAL
GALAXY

RADIO LIGHT
JET FROM
ELLIPTICAL
GALAXY NGC 6251
300 MILLION
LIGHT YEARS
AWAY

400,000
LIGHT
YEARS
LONG

THE EXPLODING GALAXY
CYGNUS "A"..
ABOUT 700 MILLION
LIGHT YEARS AWAY..
IS ALSO A TREMENDOUS
RADIOLIGHT SOURCE

THE RADIO GALAXY
CYGNUS "A".. WITH
THE OPTICAL PART
NEAR THE MIDDLE

AT THE CENTER IS A
BRILLIANT, VERY ACTIVE
INTENSE RADIO AND
INFRARED LIGHT SOURCE..
ONLY 12 LIGHT DAYS
ACROSS.

WITHIN THE OPTICAL
PART OF THE GALAXY,
ARE TWO CLOUDS..
ABOUT 20,000 LIGHT
YEARS APART...
THAT APPEAR TO
HAVE BEEN SENT IN
OPPOSITE DIRECTIONS.

IN "RADIOLIGHT"
WE SEE TWO GREAT
CLOUDS OF ENERGETIC
GAS OUT TO DISTANCES
OF 100,000 LIGHT
YEARS FROM THE CENTER

GALAXIES OF GALAXIES OF GALAXIES

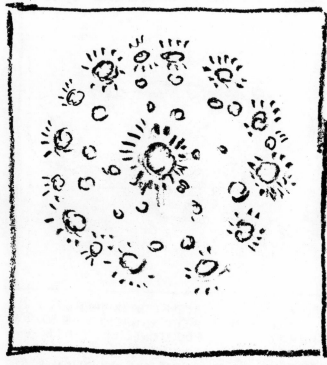

THERE ARE GALAXIES OF GALAXIES

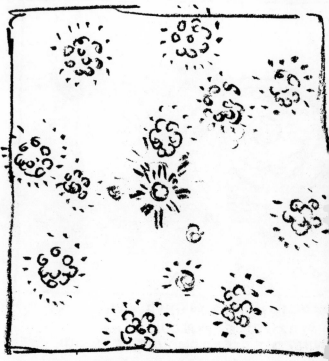

AND GALAXIES OF GALAXIES OF GALAXIES CALLED "SUPERCLUSTERS"

AS GALAXIES ARE CLASSIFIED BY SHAPE, SO ARE GALAXIES OF GALAXIES

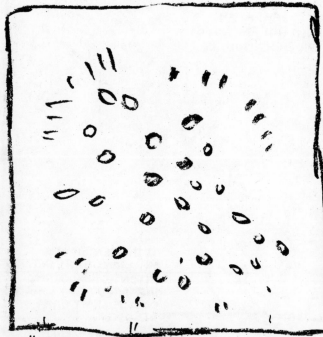

AN "IRREGULAR" CLUSTER OF GALAXIES APPEARS TO HAVE NO SPHERICAL SYMMETRY AND NO CENTRAL CONCENTRATION

A RICH IRREGULAR CLUSTER OF THOUSANDS OF GALAXIES OF ALL TYPES . . . THE CLUSTER OF GALAXIES IN VIRGO

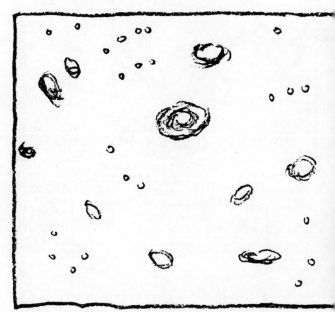

THE VIRGO CLUSTER OF GALAXIES IS AN IRREGULAR CLUSTER

OME GALAXIES MOVE
OGETHER

"RICH CLUSTERS" OF
HUNDREDS OF GALAXIES..
AND
N "GREAT CLUSTERS" OF
THOUSANDS OF GALAXIES

STRONOMERS KNOW OF
ABOUT THREE THOUSAND
GREAT" CLUSTERS OF GALAXIES

THE NEAREST "RICH" CLUSTER OF GALAXIES
IS ABOUT 60 MILLION LIGHT YEARS AWAY
IN THE CONSTELLATION VIRGO

A BIG REGULAR CLUSTER OF MORE THAN
THOUSAND GALAXIES..
CLUSTER OF GALAXIES IN COMA BERENICES

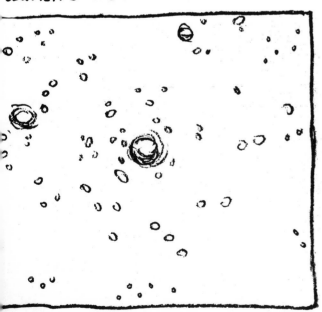

HE "GREAT" CLUSTER OF GALAXIES
N COMA BERENICES IS A REGULAR
LUSTER..

IS IS THE "GREAT" GALAXY OF GALAXIES
N CORONA BOREALIS

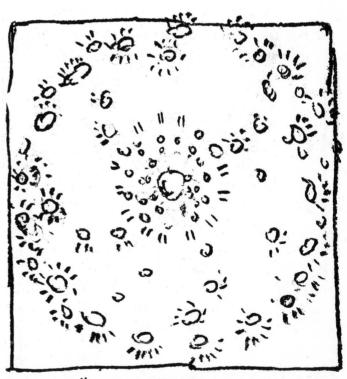

"REGULAR" GREAT CLUSTERS OF
THOUSANDS OF GALAXIES APPEAR
LIKE GREAT SPHERES IN THE HEAVENS

AMAZING "GALAXIES" OF THOUSANDS OF
GALAXIES OF TENS OF BILLIONS OF
STARS EACH.

THE CENTER OF THE GALAXIES OF GALAXIES

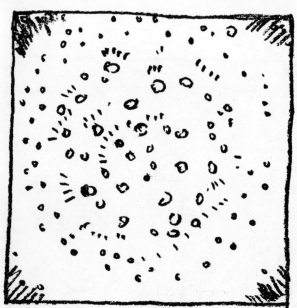

WHO KNOWS
WHAT IS AT
THE CENTER OF
GALAXIES OF
GALAXIES
?

NEAR THE CENTER OF SOME
CLUSTERS OF GALAXIES
ARE GREAT CONCENTRATIONS
OF GALAXIES... SURROUNDED
BY A GREAT HALO OF STARS

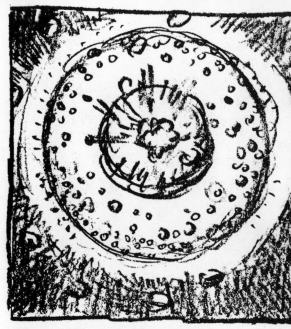

NEAR THE CENTER OF SOME
GALAXIES OF GALAXIES
ARE SUPERGIANT GALAXIES
SURROUNDED BY
A HALO OF STARS

THERE ARE
BRILLIANT
GALAXIES
NEAR THE
CENTER...

IS IT
THE
QUALITY
OF
LIGHT?

OR ARE THEY
MOVING AT
INCREDIBLE
SPEEDS?

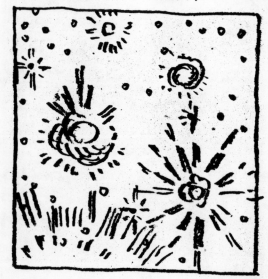

THE GALAXIES NEAR
THE CENTER
APPEAR TO BE MOVING
AT INCREDIBLE SPEEDS

THERE APPEARS TO
A GRAVITATIONAL
"MAELSTROM" AT T
CENTER OF SOME

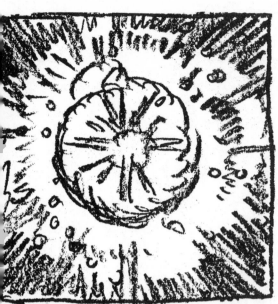

"HEAD-TAIL"
RADIO GALAXIES

SEEN IN RICH
CLUSTERS OF
GALAXIES

MIGHT BE CAUSED
BY THE INTERGALACTIC
GAS

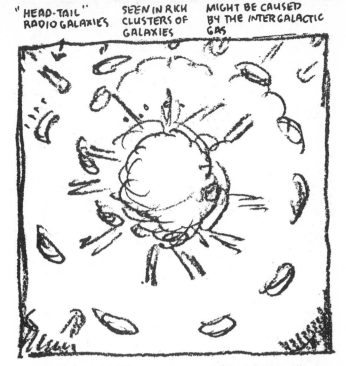

REAT" CLUSTERS OF GALAXIES
PPEAR TO BE HELD TOGETHER
Y A VERY GREAT FORCE ..
EN TIMES GREATER THAN
HAT CAN BE ACCOUNTED FOR
Y WHATS VISIBLE TO US NOW

RECENT OBSERVATIONS
IN X-RAY AND
RADIO LIGHT
SHOW EVIDENCE OF
TITANIC EXPLOSIONS
AT THE CENTER OF
SOME GALAXIES OF
GALAXIES.

IT APPEARS THAT
GREAT CLOUDS
OF HIGH ENERGY
SUB-ATOMIC PARTICLES
HAVE BEEN SENT
FORTH FROM
THE CENTER

AND THAT
THE SPACE BETWEEN
THE GALAXIES IS RICH
IN HOT INTERGALACTIC
GAS.. NOT SEEN IN
"VISIBLE LIGHT"
PHOTOGRAPHS

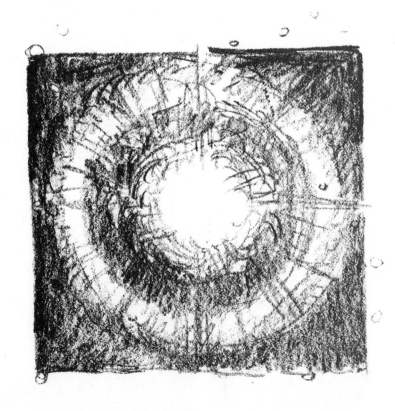

WHO KNOWS
WHAT
IS
WITHIN
?

NOBODY KNOWS HOW GALAXIES BEGIN.
THERE ARE MANY THEORIES..
BASED ON A LITTLE EVIDENCE
AND A LOT OF SPECULATION

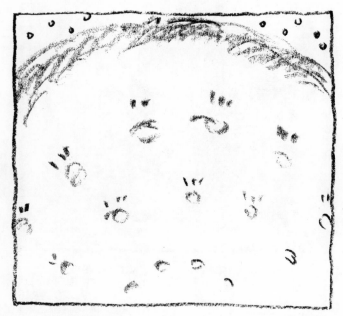 

SOME SAY
ALL THE GALAXIES
WERE FORMED
AT THE SAME TIME
BILLIONS OF YEARS AGO

FROM GREAT
CLOUDS

THAT BEGAN
WHIRLING

INTO
WHIRLING
CLOUDS
WITHIN
CLOUDS

THAT
BECAME
THE
GALAXIES
OF
GALAXIES

IN A GREAT CLUSTER OF ABOUT 10,000
GALAXIES IN BERENICE'S HAIR..

80 MILLION -- 100 MILLION LIGHT YEARS AWAY

NEAR THE CENTER OF A CLUSTER OF ABOUT
10,000 GALAXIES..
THE GREAT CLUSTER OF GALAXIES IN HERCULES
300 MILLION LIGHT YEARS AWAY

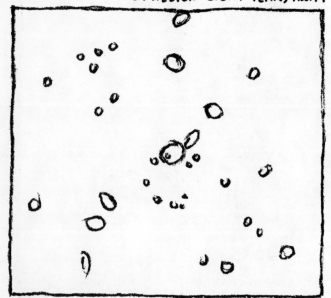 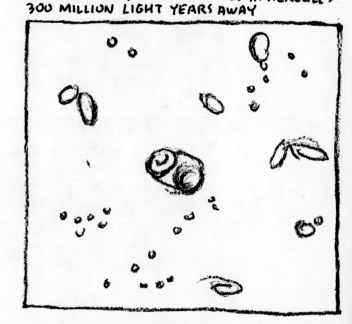

HERE ARE PICTURES
OF REGIONS
OF SOME OF THE
GALAXIES OF GALAXIES
WE CAN PHOTOGRAPH NOW

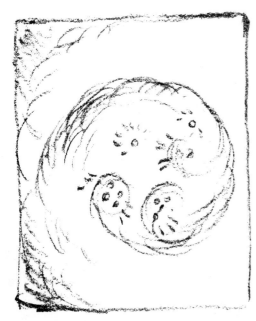

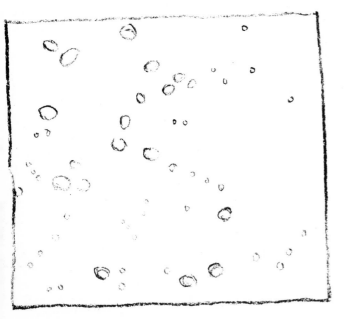

SOME SAY
GALAXIES
COME TO BE
THROUGHOUT
ALL
OF
TIME

FROM
THE
CLOUDS

OR

FROM
QUASARS

OR

FROM
"WHITEHOLES"

THE CLUSTER OF GALAXIES IN CORONA BOREALIS
800 MILLION TO 500 MILLION LIGHT YEARS AWAY

EACH OF THESE VERY FAINT DOTS OF LIGHT
ARE MILLIONS OF STARS
IN A DISTANT CLUSTER OF GALAXIES...
2 BILLION TO 5 BILLION LIGHT YEARS AWAY

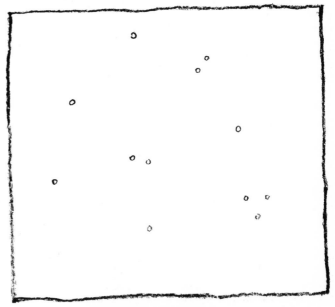

ON THE NEXT PAGES
ARE SOME OF
TODAY'S THEORIES
OF HOW GALAXIES
BEGIN:

HERE IS
ONE OF THE MOST POPULAR
OF THE
" GALAXIES FROM
THE CLOUDS "
THEORIES
OF TODAY

GREAT CLOUDS
200,000
LIGHT YEARS
ACROSS

THE
PROTOGALAXY
100,000
LIGHT YEARS
ACROSS

WHEN
THE
FIRST
STARS
LIGHT

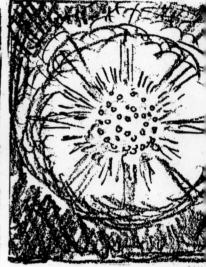

IN THE BEGINNING
OF THE GALAXIES..
THERE WERE GREAT
CLUSTERS OF CLOUDS
MOVING TOGETHER
THROUGH THE SKIES

OVER TIME, THE CLOUDS
MOVED CLOSER AND
CLOSER TOGETHER
AND EVENTUALLY
MERGED INTO A
GREAT PROTO GALAXY

THE PROTOGALAXY
MOVES INWARD...
AND INWARD

AFTER ABOUT
200 MILLION YEARS,

THE FIRST STARS
FLASH FORTH
AND LIGHT
THE SKIES
IN A BRILLIANT
BURST

LIKE THE
GIANT ELLIPTICAL
GALAXY M87
IN VIRGO

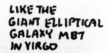

AND,
AS THE COSMIC GAS
IS ALL USED UP,
NO MORE
NEW STARS
COME TO BE..

THE BRIGHTER
STARS,
OVER THE YEARS,
BECOME THE
OLDER, DIMMER
STARS

THE OLDER STARS
LIVE ON
IN A GIANT
SPHERICAL
FAMILY

ACCORDING TO SOME,
SPIRAL GALAXIES
BEGAN IN THE SAME
WAY

IN THE EARLY YEARS,
LIGHTING MANY
GENERATIONS
AND GREAT NUMBERS
OF STARS.

THE CLOUD IS
INCREDIBLY DENSE
AT THE CENTER

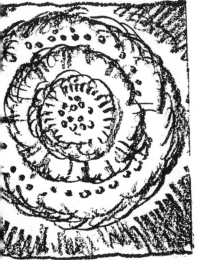
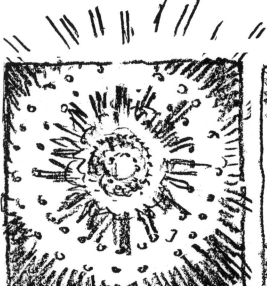
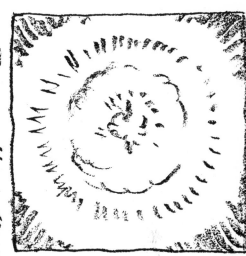

...ER THE YEARS, THE
...RST BURST OF STARS
...EAVES THE SKIES
...ND
...HE PROTOGALAXY
...ONTINUES ITS
...NWARD JOURNEY
...OYING FASTER
...ND FASTER

MOVING FURTHER
AND FURTHER INWARD..
LIGHTING
GENERATION
AFTER
GENERATION
OF STARS

UNTIL..
IN A NUCLEUS ABOUT
10,000 LIGHT YEARS
ACROSS.. AFTER
ABOUT 300 MILLION
YEARS..
THE STARS LIGHT
AT A FANTASTIC RATE.

AND
THE GALAXY LIGHTS UP
INTO ONE OF THE
BRIGHTEST OBJECTS
IN THE HEAVENS
OF ALL TIME.

MANY ANCIENT STARS
ARE SPHERICALLY
DISTRIBUTED ABOUT
THE CENTER OF
SPIRAL GALAXIES

...FTER WHICH
...HE COSMIC CLOUDS
...ETTLED
...NTO A
...REAT DISC...

ROTATING OUTWARD
TO ORBIT
THE CENTER OF
THE GALAXY

TO LIGHT
GENERATIONS
AND GENERATIONS
OF STARS
IN THE RICH CLOUDS
OF THE
SPIRAL ARMS

AS THE RICH CLOUDS
MOVE THROUGH THE
WAVES AND VALLEYS
OF THE SPIRAL..
BRILLIANT NEW
BLUE STARS
FLASH FORTH.

TO RIDE THE WAVES
OF THE OUTER REALMS
EVENTUALLY- AFTER
MILLIONS OF YEARS
TO COOL DOWN
INTO BRIGHT
YELLOW STARS
LIKE THE SUN

FOR THOSE WHO VIEW
THE GALAXIES AS
HAVING COME FROM
THE CLOUDS..

THERE ARE A LOT OF
THEORIES OF HOW
THE DIFFERENT KINDS
OF GALAXIES MIGHT
HAVE COME TO BE..

EACH IN THEIR OWN WAY
WITHOUT CHANGING
FROM ONE INTO ANOTHER

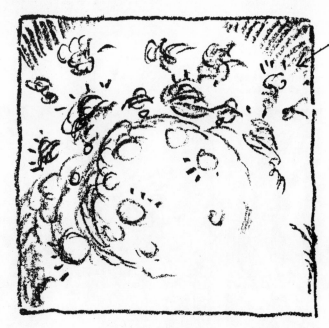

THE
IRREGULARS
WHIRLING
THE FASTEST

SOME SAY THE
"SPEED OF WHIRLING"
OF EACH PROTOGALAXY
DETERMINES THE
SHAPE OF THE GALAXY
WE SEE TODAY

THE MOST SPHERICAL
GALAXIES HAVING
DEVELOPED FROM
THE PROTOGALAXIES
THAT WHIRLED
THE SLOWEST..

FROM THE ELLIPTICALS,
THROUGH THE SPIRALS,
UP TO THE "IRREGULAR"
GALAXIES.. THAT MAY
HAVE COME FROM THOSE
THAT WHIRLED THE
FASTEST.

THE DENSITY OF
THE ORIGINAL CLOUD
ACCORDING TO SOME..

THE SPEED OF TRAVEL,
ACCORDING TO SOME...

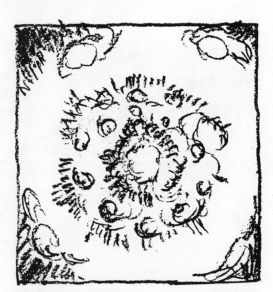

ONE VIEW TELLS OF
THE SHAPES OF GALAXIES RESULTING
FROM THE DENSITY OF THE CLOUD
FROM WHICH THE PROTOGALAXIES
ARE BORN..

THE MOST SPHERICAL, EARLIEST LIT,
MOST UNIFORMLY LIT GALAXIES
COMING FROM THE HIGHEST DENSITY
AT THE CENTER.

ANOTHER VIEW
TELLS OF GALACTIC SHAPES
THAT ARE DETERMINED BY
THE SPEED AT WHICH THEY
MOVE THROUGH THE INTERGALACTIC
MEDIUM.

THE MOST SPHEROIDAL,
ALMOST GASLESS SPIRAL GALAXIES
HAVING BEEN BLOWN CLEAN
IN A GREAT INTERGALACTIC WIND.

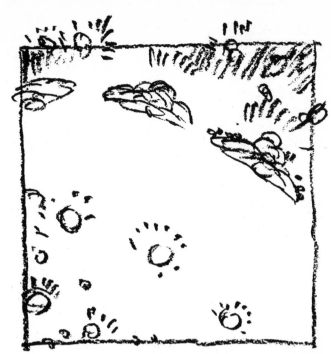

SOME SAY THE
RATE AT WHICH THEY
USE UP THEIR GAS
AND DUST DETERMINES
THE SHAPE OF THE
GALAXY WE SEE TODAY

THAT ALL GALAXIES
WERE FORMED AT THE SAME TIME
BILLIONS OF YEARS AGO...
THE ELLIPTICALS, PROCESSING THEIR
GAS AND DUST THE QUICKEST,
COMPLETED THEIR STAR MAKING
AFTER A FEW GENERATIONS...

THE SPIRALS, RICH IN GAS AND DUST
ARE STILL LIGHTING UP STARS
AFTER MANY GENERATIONS
IN THEIR CLOUD LIKE ARMS..
AND THE IRREGULAR GALAXIES
LIGHT UP STARS IN BURSTS
EVERY NOW AND THEN.

SOME SAY
THAT SPIRAL GALAXIES COME TO BE
FROM CLOSE ENCOUNTERS BETWEEN
GALAXIES PASSING IN SPACE..
CREATING GALACTIC TIDES
THAT BECOME THE SPIRAL DENSITY
WAVES OF THE
GALAXIES.

ANOTHER VIEW TELLS OF

GALAXIES PASSING THROUGH
ONE ANOTHER.. THE STARS SO FAR
APART THERE WOULD BE NO COLLISSIONS.
AFTER WHICH, THE SHAPES OF
THE GALAXIES WOULD BE
QUITE DIFFERENT THAN
BEFORE.

SOME SAY THAT
GALAXIES, OVER TIME,
CHANGE FROM ONE KIND
TO ANOTHER

-
SOME SAY GALAXIES
COME FROM
THE CLOUDS

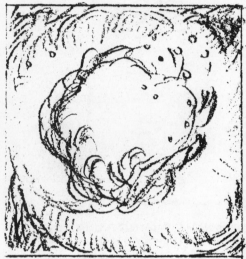  

THE WHIRLING CLOUDS
MOVE INWARD TO BECOME
A PROTOGALAXY..

MOVING FURTHER
INWARD TO BEGIN
TO LIGHT
THE STARS

FIRST.. LIGHTING STARS
IRREGULARLY..
THE GREATEST
NUMBER OF STARS
LIGHTING THE CENTER
WHERE THERE IS MORE
DUST AND GAS

THEN.. SUPERNOVA EXPLOSIONS
AT THE CENTER SEND
GREAT WAVES THAT BEGIN
A GREAT VARIETY OF STARS
THAT LIGHT UP AS THE
ROTATING SPIRAL
"ARMS" OF A
SPIRAL GALAXY

OVER THE YEARS..
THE SPIRAL ARMS
BECOME MORE
TIGHTLY WOUND..
UNTIL THE ARMS
DISAPPEAR AND
AN ELLIPTICAL GALAXY
REMAINS

FROM A COMPLETELY
DIFFERENT POINT OF
VIEW..

SOME SAY THE GALAXIES
COME DIRECTLY FROM A
"SINGULARITY"
OR FROM A
"WHITEHOLE"

THE DIFFERENT
KINDS OF GALAXIES
ARE "PHASES"
OF A COSMIC
SINGULARITY

WHICH FIRST
COMES TO LIGHT
AS A WHITEHOLE

AND THEN
COMES TO LIGHT
AS A QUASAR

AND THEN
COMES TO LIGHT

AS A "BLAZAR"
ONE OF THE
BRIGHTEST
LIGHTS IN
ALL OF TIME

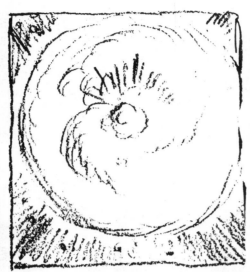

ANOTHER VIEW OF THE GALAXIES FROM THE CLOUDS TELLS OF THE WHIRLING CLOUDS MOVING INWARD TO FORM A QUASAR ' ' ' ' ' '

WHICH, OVER TIME, CHANGES INTO A "BLAZAR".. AND THEN INTO THE VARIOUS TYPES OF GALAXIES ..

A SEYFERT GALAXY.. AND THEN..

AN ELLIPTICAL GALAXY.. AND THEN..

A PECULIAR GALAXY.. AND THEN ..

A SPIRAL GALAXY

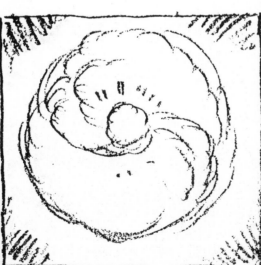

AND THEN COMES TO LIGHT AS A "SEYFERT"

THEN · AS AN "N·TYPE" GALAXY .

·· AN ELLIPTICAL WITH A VERY BRIGHT CENTER ··

AND.. OVER THE YEARS.. AS THE QUASAR QUIETS DOWN

THE GALACTIC DISC MOVES OUTWARD INTO A SPIRAL GALAXY

THE CENTER OF THE GALAXY

THE CENTER OF OUR GALAXY
IS HIDDEN FROM VIEW
BY THE GREAT CLOUDS
OF SAGITTARIUS

A RADIOLIGHT
MAP OF THE
CENTER
GIVES US
THE BEGINNING
OF A PICTURE
OF THE CENTER

THE
GALACTIC
EQUATOR

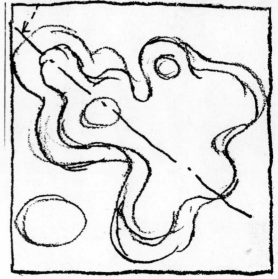

THERE'S
A TREMENDOUS SOURCE
OF ENERGY
AT THE CENTER.
NOBODY KNOWS
WHAT IT IS.

IN THE 1930's,
THE FIRST
RADIOLIGHT
ASTRONOMER
OBSERVED
A TREMENDOUS
RADIOLIGHT
SOURCE
BEYOND THE
CLOUDS OF
SAGITTARIUS

IT'S NOW
THOUGHT TO BE
THE CENTER
OF OUR
GALAXY...
ASTRONOMERS
OF TODAY
ARE LOOKING
TO THE CENTER
IN ALL THE
WAVELENGTHS

TRYING TO
UNDERSTAND
WHAT COULD
BE CREATING
SO
MUCH
"LIGHT"

THE INFRARED ENERGY
AT AND NEAR THE CENTER
IS ALMOST AS GREAT AS
THE ENERGY OF ALL THE
VISIBLE LIGHT OF
THE ENTIRE GALAXY

IN INFRARED LIGHT,
THE CENTER LOOKS LIKE
A SEYFERT GALAXY,
ONLY NOT AS INTENSE.

WHO KNOWS
WHATS AT
THE CENTER?

THOUSANDS
OF BRILLIANT
STARS
?

THOUSANDS
OF FLASHING
PULSARS
?

A
SUPERNOVA
HEAVEN
?

ONE
INCREDIBLE
GREAT
SUPER
STAR
?

A POWERFUL
RADIO SOURCE...
"SAGITTARIUS 'A'"
LIES NEAR
THE CENTER

IN THE GALACTIC DISC,
HYDROGEN APPEARS
TO PLOW OUTWARD
FROM THE CENTER...
AT SPEEDS UP TO
100 MILES A SECOND!

CONTINUING OUTWARD
INTO THE SPIRAL ARMS...
(TO, PERHAPS...SOMEDAY
RISE INTO THE GALACTIC
HALO AND EVENTUALLY
RETURN TO THE CENTER?)

THERE APPEAR TO BE
POWERFUL GRAVITATIONAL
WAVES COMING FROM
THE CENTER.

GRAVITATIONAL WAVES
ARE "RIPPLES" OF
SPACE-TIME ITSELF
TINY TINY WAVES
MOVING AT
THE SPEED OF LIGHT

THE SOURCE WOULD
HAVE TO BE INCREDIBLY
POWERFUL TO SEND
SUCH TINY WAVES
SUCH A DISTANCE

A
WHITEHOLE
LIGHTHOLE
?

A
SUPERMASSIVE
BLACK
HOLE
?

A
QUASAR
?

THOUSANDS
OF
WHITEHOLES
?

OR
COMBINATIONS
OF THE
ABOVE
?

OR
NONE
OF
THE ABOVE
?

WHAT'S AT THE CENTER
OF OUR OWN GALAXY?
HERE ARE SOME OF
THE LATEST IDEAS...

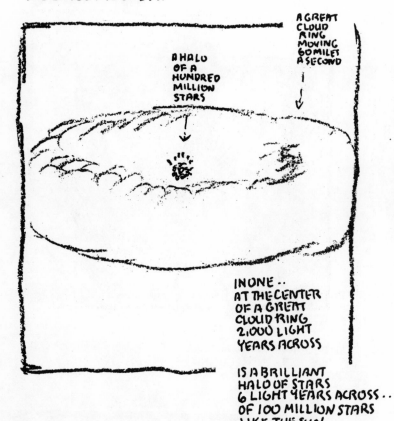

A GREAT
CLOUD
RING
MOVING
60 MILES
A SECOND

A HALO
OF A
HUNDRED
MILLION
STARS

IN ONE..
AT THE CENTER
OF A GREAT
CLOUD RING
2,000 LIGHT
YEARS ACROSS

IS A BRILLIANT
HALO OF STARS
6 LIGHT YEARS ACROSS..
OF 100 MILLION STARS
LIKE THE SUN

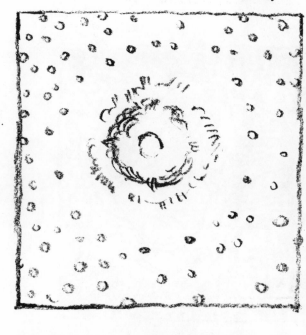

A
HUNDRED
MILLION
STARS

WITHIN THE STARS
IS A CENTER
LIKE A SEYFERT CENTER
(ONLY NOT AS
POWERFUL)

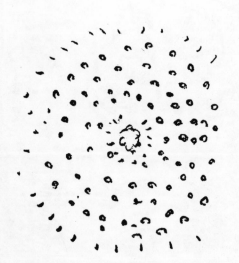

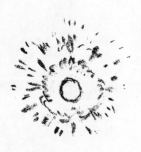

IN OTHER
PICTURES..
DEEP WITHIN
A BRILLIANT
FIELD OF STARS..
IS THE
CENTER

IN ONE..
THE CENTER IS
ONE
SUPERMASSIVE
SPINNING
STAR

IN ANOTHER..
ITS
THOUSANDS
OF
SPINNING
PULSARS

THAT CAME
TO BE FROM AN
INCREDIBLE
BURST OF
THOUSANDS
OF
SUPERNOVAS

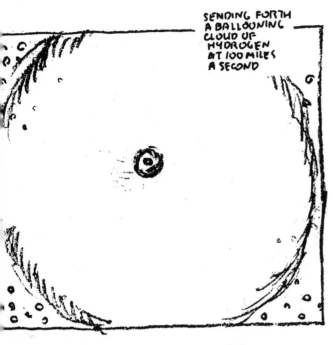

SENDING FORTH
A BALLOONING
CLOUD OF
HYDROGEN
AT 100 MILES
A SECOND

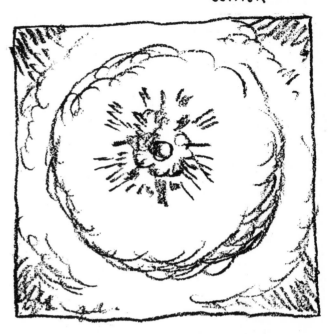

ZOOM IN
TO THE
CENTER

IN ANOTHER
PICTURE · ·
AT THE CENTER
OF A LIGHT GAS
CLOUD SPHERE
HUNDREDS OF
LIGHT YEARS
ACROSS

IS A
DENSE
CLOUD

ONE
LIGHT YEAR
ACROSS

AND
AT THE CENTER
OF THE CLOUD
WITHIN THE CLOUD
IS A
QUASAR

IN SOME PICTURES · ·
THERE IS A
SUPERMASSIVE
BLACKHOLE

IN SOME · ·
THERE IS
A GREAT
WHITE HOLE

IN SOME ‒
THERE IS
A
SINGULARITY

QUASARS

THE FIRST QUASARS
WERE FOUND
IN 1961

NO ONE KNOWS
WHAT QUASARS ARE

THEY ARE STAR LIKE
IN APPEARANCE

THEY SEEM TO BE
AMAZINGLY
POWERFUL

THERE ARE SEVERAL HUNDRED
KNOWN "QUASARS" AND "QSOs"

ITS THOUGHT
THAT THEY
LIGHT UP
FOR ABOUT
100 MILLION YEARS .

GIVING OFF AS MUCH
LIGHT IN THAT TIME
AS AN ENTIRE
GALAXY DOES IN
15 BILLION YEARS

IF THE POPULAR
"COSMOLOGICAL"
INTERPRETATION
OF THE REDSHIFTS
IS RIGHT . .
THEY ARE
AMAZINGLY
BRILLIANT ENERGY
SOURCES . .

50 TO 100 TIMES
MORE BRILLIANT
THAN AN ENTIRE
GALAXY OF
100 BILLION
STARS !

AND THEY'RE
BILLIONS OF LIGHT
YEARS AWAY
AND ZOOMING
AWAY FROM US
AT INCREDIBLE
SPEEDS !

THERE APPEARS
TO BE A HUGE
"BLOB" SPEEDING
AWAY FROM IT
AT NEARLY
THE SPEED OF
LIGHT .

A RECENT
OBSERVATION
SUGGESTS
SOMETHING
NEAR 3C273 . .
MOVING
AMAZINGLY . .
AT TEN TIMES
THE SPEED
OF LIGHT ! !
? ? ?

THE QUASAR 3C273
IN VIRGO
COULD BE THE
NEAREST QUASAR . .
ABOUT 3 BILLION
LIGHT YEARS AWAY

ITS 100 TIMES
BRIGHTER
THAN THE BRIGHTEST
ORDINARY GALAXY . . .

100,000 TIMES
BRIGHTER IN
INFRARED LIGHT
THAN THE
ENTIRE
MILKY WAY
GALAXY

IT HARBORS A
MYSTERIOUS
LUMINOUS JET
ABOUT 150,000
LIGHT YEARS ACROSS . .

IT HAS 5 RADIO
COMPONENTS
IN ADDITION
TO THE
OPTICAL
SOURCE

RADIOLIGHT PHOTONS
ARISE FROM TREMENDOUS
ENERGY SOURCES
THAT SEND ELECTRONS
TRAVELING AT SPEEDS
NEAR THAT OF LIGHT.!

SOME
QUASARS
FLARE
UP
EVERY
NOW
AND
THEN

QSOs
(QUASI-STELLAR
OBJECTS)
EMIT
INTENSE
VISIBLE LIGHT..
BUT NOT MUCH
RADIOLIGHT.

SOME ARE
ALSO
BRIGHTLY
ULTRAVIOLET
AND
INFRARED

QUASARS
(QUASI-STELLAR
RADIO OBJECTS)
EMIT
INTENSE
RADIO LIGHT
AND
VISIBLE LIGHT..

SOME ARE
ALSO
BRIGHTLY
ULTRAVIOLET
AND
INFRARED

THE QUASAR 3C345
FLARES UP ABOUT
ONCE A YEAR
WITH THE POWER OF
THOUSANDS OF BILLIONS
OF STARS

THE QUASAR 3C279
IS THOUGHT TO HAVE
(FOR ABOUT 2 WEEKS
IN APRIL OF 1937)..
FLARED UP AS
BRILLIANT AS
TENS OF THOUSANDS
OF GALAXIES.

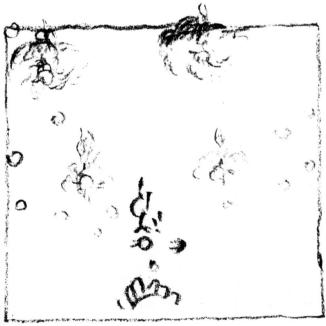

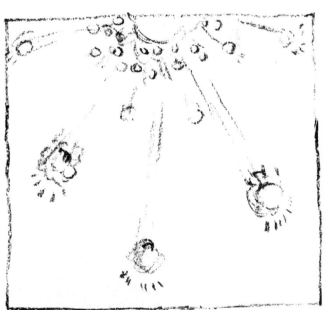

SOME SAY
THAT QUASARS
MIGHT COME
INTO BEING
EVERY NOW
AND THEN
AS THE
BEGINNING
OF
GALAXIES

OR
THAT THEY MIGHT
COME TO BE
OUT OF THE
CLOUDS..
JUST AS GALAXIES
MIGHT..

OR THAT THEY
MIGHT BE
SENT FORTH
OUT OF
EXISTING
GALAXIES..

SOME SAY THAT QUASARS
MAY HAVE EXISTED
FROM THE "BEGINNING"

THAT THEY MIGHT BE
MATTER POPPING INTO
THE UNIVERSE FROM
THE COLLAPSE OF
ANOTHER..
EACH QUASAR
BEING A
"LITTLE BANG"

SOME SAY
THAT THEY MIGHT
COME TO BE
THROUGHOUT
THE UNIVERSE
OF ALL TIME..

OUT OF
WHITEHOLE
SINGULARITIES

HERE ARE SOME OF
THE LATEST IDEAS
OF HOW QUASARS
MIGHT HAVE COME
TO BE

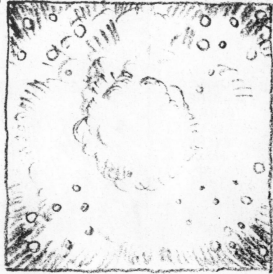

SOME SAY
THE QUASAR
MAY BE A
"LIGHTING UP"
PHASE OF
THE CENTER
OF A GALAXY

PERHAPS, IN THE
FINAL INWARD
FALL.. SOON AFTER
THE GALAXY IS
BORN..
MANY STARS
COLLIDE AT
THE CENTER
TO LIGHT AS
A QUASAR

OR, PERHAPS,
IN THE FINAL
INWARD FALL,
THE
REMAINING
CLOUDS
COME UNDER
SUCH GREAT
PRESSURE
THEY LIGHT
AS A QUASAR

OR, PERHAPS,
A BLACKHOLE
SWALLOWS IN
NEARBY STARS..
GETS BIGGER
AND BIGGER
AND BIGGER..
AND "LIGHTS"
AT THE CENTER

HOWEVER
IT MIGHT HAPPEN,
THE CENTER
"BRIEFLY"
LIGHTS UP
BRIGHTER
THAN THE
ENTIRE
GALAXY

AND..
WHEN THE
CENTER
TURNS OFF
OPTICALLY,
THE
SURROUNDING
GALAXY IS
OBSERVABLE

SOME SAY THAT
QUASARS ARE
INCREDIBLY
BRIGHT VERSIONS
OF NORMAL
GALACTIC CENTERS..
UP TO A THOUSAND
TIMES BRIGHTER

SOME QUASARS
HAVE A HAZY
APPEARANCE,
AS THOUGH THEY
MAY BE BEGINNING
TO CHANGE INTO
A GALAXY

PERHAPS QUASARS
ARE VERY YOUNG
GALAXIES..
COMING TO LIGHT
WITH VERY
BRIGHT
"NUCLEUS"
CENTERS

PERHAPS
THEY COME
TO BE OUT OF
"WHITE HOLE
SINGULARITIES"

FROM
BEYOND
TIME SPACE
AS WE KNOW
IT

"BLAZARS"
COULD BE
PHASES OF
QUASARS

"BLAZARS"
(BL LACERTAE
OBJECTS)
HAVE ALSO BEEN
FOUND AT THE
CENTER OF
GALAXIES

BLAZARS
HAVE A SIMILAR
SPECTRUM TO
QUASARS..

THEY COULD BE
ROTATING
VERY RAPIDLY,
JUST AS
QUASARS
COULD BE.

SOMETIMES
THEY'RE
NOT AS BRIGHT
AS QUASARS

SOMETIMES
THEY'RE
A LOT
BRIGHTER

SOMETIMES
THEY BLAZE UP
TO BE THE
BRIGHTEST
OBJECT IN
THE SKY

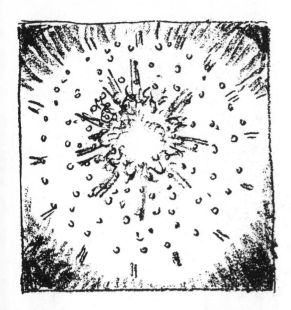

SOME SAY
THE QUASAR
BEGINS AS MILLIONS
OF STARS . .
SO CLOSE TOGETHER
THAT NEUTRON STARS
SPIRAL
RIGHT INTO
THE MAIN STARS

CREATING
UNSTABLE SYSTEMS
WHICH SOON
EXPLODE . .
ITS "BRAND X"
SUPERNOVAS . .
AN INCREDIBLE
ENERGY . .
APPEARING
AS A QUASAR .

THE "SPINAR" THEORY
DESCRIBES A GIGANTIC
"PULSAR" LIKE QUASAR . .

A SPINNING STAR
100 MILLION TIMES AS
MASSIVE AS THE SUN
CALLED A "SPINAR"

AS IT LOSES ENERGY
IT CONTRACTS
TO SPIN EVEN
FASTER AND FASTER .

AND BECOMES
BRIGHTER AND
BRIGHTER . .

THE ROTATIONAL ENERGY
OF THE "SPINAR"
IS CONVERTED INTO
TREMENDOUS ENERGY
THAT RADIATES
THE LIGHT OF A
QUASAR

PERHAPS,
AFTER THE
QUASAR
AND
BLAZAR
PHASES . .

THE
"SEYFERT"
COMES
TO LIGHT

THE REDSHIFTS
OF SEYFERT
GALAXIES
ARE OFTEN
SOMEWHERE
BETWEEN THOSE
OF QUASARS
AND REGULAR
GALAXIES

THE CENTERS OF
SEYFERT GALAXIES
EMIT TREMENDOUS
AMOUNTS OF
INFRARED ENERGY

. . ALMOST EQUAL
TO THE TOTAL
VISIBLE LIGHT
ENERGY OF THE
ENTIRE GALAXY .

. . AND RADIO LIGHT .

THE CENTERS OF
SEYFERT GALAXIES
SEND FORTH
GREAT CLOUDS
OF GAS . .

IS THIS
THE BEGINNING
OF THE CLOUDS
OF THE STARS
OF THE GALAXIES

?

ARE GALAXIES AND QUASARS
CONNECTED IN SOME WAY
NOT YET OBSERVABLE ?

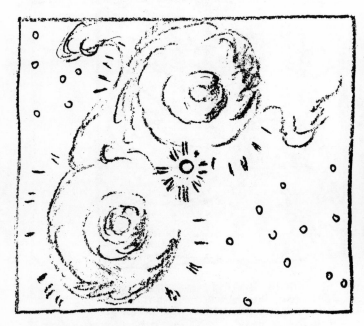

THE "OPTICAL" QUASAR 3C-47
LIES NEAR THE MIDDLE OF
TWO RADIO EMITTING REGIONS
700,000 LIGHT YEARS APART

QUASARS
HAVE BEEN
DISCOVERED

NEAR MANY
SEYFERT
GALAXIES

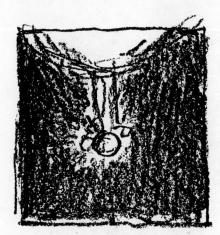

THE GALAXY NGC 4319
AND THE QUASAR
MARKARIAN 205..
VERY DIFFERENT
IN REDSHIFTS..
APPEAR TO BE
CONNECTED BY
A "BRIDGE"
OF GAS

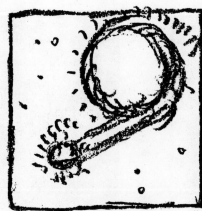

THE SEYFERT GALAXY
NGC 7603
AND ITS PECULIAR
COMPANION GALAXY
APPEAR TO BE
CONNECTED BY
LUMINOUS
FILAMENTS

SOME SAY THAT
GALAXIES
MIGHT BE
SENT FORTH
FROM GALAXIES

NEAR
EXPLODING
GALAXIES

NEAR
PECULIAR
GALAXIES

SOME SAY THAT
GALAXIES
MIGHT BE
SENT FORTH
FROM
QUASARS

SOME SAY THAT
QUASARS
MIGHT BE
SENT FORTH
FROM
GALAXIES

GALAXIES
MIGHT
APPEAR
OUT OF THE
LIGHT OF
THE QUASAR

IN RECENT YEARS
ASTRONOMERS
HAVE OBSERVED
A GREAT VARIETY
OF APPARENT ASSOCIATIONS
BETWEEN
GALAXIES AND GALAXIES
QUASARS AND GALAXIES
AND
QUASARS AND QUASARS
THAT CAN'T BE EXPLAINED
IN TERMS OF THE
HUBBLE LAW INTERPRETATION
OF THE LIGHT

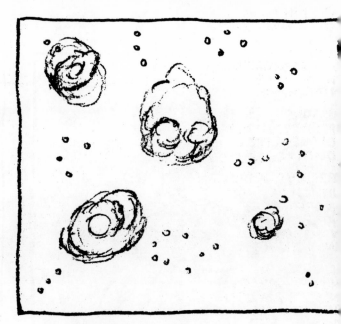

IN "THE STEPHAN'S QUINTENT OF GALAXIES"
IN PEGASUS

THE FOUR LITTLER GALAXIES
SHOW DRAMATICALLY GREATER
REDSHIFTS THAN THE BIGGEST
GALAXY OF THE FIVE

OPTICALLY DIM GALAXIES
WITH LARGE REDSHIFTS
HAVE BEEN FOUND VERY NEAR
SOME PECULIAR GALAXIES
AND ON BOTH SIDES OF
SOME PECULIAR GALAXIES

NEAR THE BRILLIANT
"EXPLODING" GALAXY NGC 520

THERE APPEAR TO BE
FOUR QUASARS
"IN A ROW"

THERE ARE FOUR
LITTLE COMPANION
GALAXIES
AROUND
THE BARRED SPIRAL
GALAXY 2859
IN LEO MINOR.

QUASARS
APPEAR TO LIE NEAR
OF THE 4 LITTLE
COMPANION GALAXIES.

ARE THEY ASSOCIATED
WITH THEM ? . .
OR ARE THEY
FAR, FAR AWAY,
IN THE SAME
LINE OF SIGHT ?

ACCORDING TO
THE "COSMOLOGICAL"
INTERPRETATION OF
THE REDSHIFTS,
THE GALAXIES
ARE ABOUT
70 MILLION LIGHT
YEARS AWAY . .
AND THE QUASARS
ARE BILLIONS OF
LIGHT YEARS AWAY

THE BARRED
SPIRAL
NGC 1073
IN CETUS

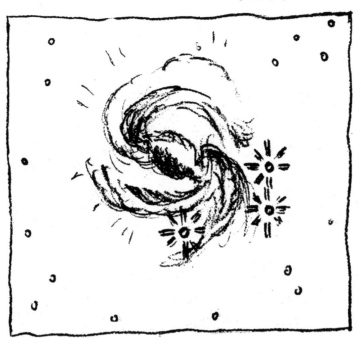

THREE QUASARS
APPEAR TO LIE NEAR
THE BARRED SPIRAL
GALAXY NGC 1073
IN CETUS .

ARE THEY ASSOCIATED
WITH THE GALAXY . .
OR ARE THEY
FAR, FAR AWAY
IN THE SAME
LINE OF SIGHT ?

ACCORDING TO
THE "COSMOLOGICAL"
INTERPRETATION
OF THE REDSHIFTS

THE GALAXY IS
ABOUT 75 MILLION
LIGHT YEARS AWAY . .
AND THE QUASARS
ARE 5 BILLION,
8 BILLION AND
9 BILLION LIGHT
YEARS AWAY !

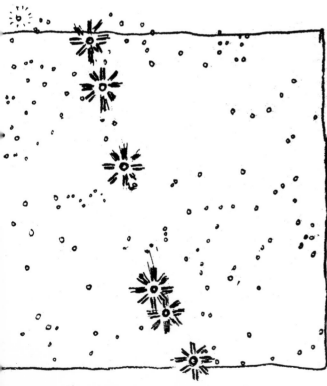

HERE ARE SIX QUASARS IN A ROW
LITTLE TO THE EAST OF
THE CONSTELLATION LEO

SOME THINK THEY MIGHT HAVE BEEN
SENT FORTH IN PAIRS
FROM A CENTRAL UNSEEN SOURCE !

MOST DISTANCES AND ENERGIES IN SPACE ARE DETERMINED IN ACCORDANCE WITH THE "HUBBLE LAW"

IT IMPLIES THE UNIVERSE IS EXPANDING.. IT'S BY FAR THE MOST POPULAR VIEW OF TODAY.

IT'S CALLED A "COSMOLOGICAL REDSHIFT"

A "DOPPLER" TYPE VIEW WHEREIN THE WAVES FROM A DISTANT OBJECT WOULD APPEAR "REDDER", WITH LONGER, MORE STRETCHED OUT WAVES.. THE FASTER AN OBJECT IS MOVING AWAY FROM US.

MOST ASTRONOMERS BELIEVE THAT THE HIGHER THE OBSERVED REDSHIFT OF THE QUASAR, THE FURTHER AWAY IT IS IN SPACE.

THIS IS BASED ON THE "HUBBLE LAW" INTERPRETATION OF THE REDSHIFTS.

QUASARS WITH ENORMOUS REDSHIFTS SHOULD BE, ACCORDING TO THE HUBBLE LAW, ZOOMING AWAY FROM US AT INCREDIBLE SPEEDS.

.. AT SPEEDS APPROACHING THAT OF LIGHT- NEAR THE LIMITS OF THE VISIBLE UNIVERSE

IF THE OBSERVED REDSHIFTS OF QUASARS ARE DUE TO "SOMETHING ELSE"

QUASARS COULD BE A LOT CLOSER THAN NOW THOUGHT TO BE ..

NOT AT THE "EDGE" OF THE UNIVERSE (IF THERE IS ONE)

AND NOT AS POWERFUL AS NOW THOUGHT TO BE

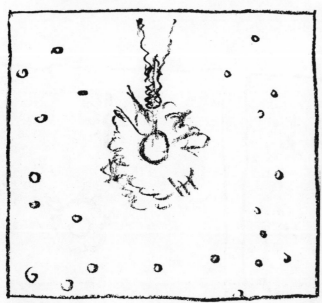

SOME SAY THAT QUASAR LIGHT IS IN A "GRAVITATIONAL" REDSHIFT.. ..WHERE, ALTHOUGH STILL TRAVELING AT THE SPEED OF LIGHT...

IN PULLING AWAY FROM THE INTENSE GRAVITATIONAL FIELD OF THE QUASAR.. IT LOSES ENERGY AND INCREASES IN WAVELENGTH!

SOME SAY THAT LIGHT INCREASES A LITTLE IN WAVELENGTH AS IT TRAVELS THROUGH TIME AND SPACE!

SOME SAY THAT QUASARS SEND FORTH "NEW MATTER" THAT INCREASES IN MASS AND WAVELENGTH AS IT AGES!

.. THAT REDSHIFT MAY BE MEASURE OF THE RELATIVE AGE OF MATTER

AS IN EVERYDAY LIFE, DIFFERENT PEOPLE CAN LOOK AT THE SAME THING AND INTERPRET IT IN ENTIRELY DIFFERENT WAYS...

DIFFERENT ASTRONOMERS CAN LOOK AT THE SAME THING AND INTERPRET IT IN ENTIRELY DIFFERENT WAYS

SOME SAY THAT IF THE QUASARS WERE "EJECTED" FROM A SOURCE SUCH AS AN EXPLODING GALAXY... OR WHATEVER..

THEY COULD BE TRAVELING AT FANTASTIC SPEEDS... THAT WOULD ACCOUNT FOR THE VERY HIGH REDSHIFTS

SOME SAY THE MYSTERY OF THE QUASARS RESTS UPON THE INTERPRETATION OF THEIR LIGHT

THAT THERE COULD BE OTHER REASONS FOR THE OBSERVED REDSHIFTS..

IF SO.. THEY NEED NOT BE BILLIONS OF LIGHT YEARS AWAY.. BUT COULD BE A LOT CLOSER TO US!

FASTER THAN LIGHT?

THE IDEA OF ANYTHING TRAVELING FASTER THAN LIGHT IS GENERALLY THOUGHT TO BE IMPOSSIBLE..

YET, IN THE VIEW OF SOME ASTRONOMERS...

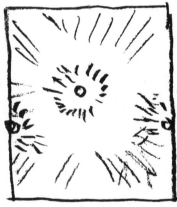 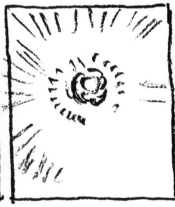 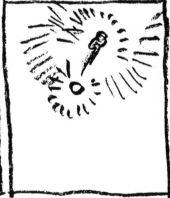

THREE QUASARS HAVE RECENTLY BEEN DISCOVERED THAT APPEAR TO BE ZOOMING AWAY FASTER THAN LIGHT!

AND, WHAT'S MORE AMAZING.. A GALAXY HAS BEEN DISCOVERED THAT APPEARS TO BE ZOOMING AWAY FASTER THAN LIGHT!

IS THE SPEED OF LIGHT NOT A LIMIT? OR IS SOMETHING ELSE GOING ON?

WAY OF SCIENCE [T]O PROPOSE A THEORY [T]O OBSERVE NATURE [T]O SUGGEST HOW THE [OBS]ERVATIONS "FIT" [WIT]H THE THEORY.

IF MANY OBSERVATIONS CONFIRM THE THEORY.. IT BECOMES ACCEPTED AS SCIENTIFIC "FACT"

IF MANY OBSERVATIONS DON'T SEEM TO CONFIRM THE THEORY.. VARIATIONS OF THE THEORY ARE PRESENTED TO TAKE THE NEW OBSERVATIONS INTO ACCOUNT

IF THIS DOESN'T WORK.. IT'S TIME FOR A NEW THEORY TO COME FORTH TO EXPLAIN BOTH THE NEW AND THE OLD OBSERVATIONS

THE SINGULARITY

AT THE
SINGULARITY
THERE IS
INFINITE
CURVATURE ..
AND "SPACE" AND
"TIME"
HAVE NO
MEANING

AT THE CENTER
OF EVERY
BLACK/WHITE
HOLE
IS A
"SINGULARITY"

.. IT'S BEYOND
DESCRIPTION
IN TERMS OF
THE SCIENCE
OF TODAY

A SINGULARITY.
ACCORDING TO
THE "RANDOMOCITY
PRINCIPLE"
LOOKS LIKE
A
"WHITEHOLE"

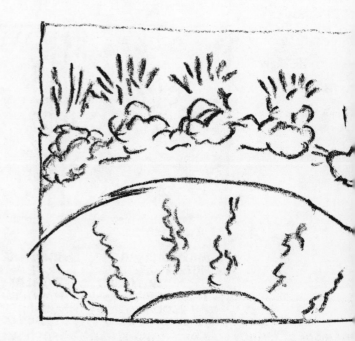

SOME SAY
THERE ARE
APPEARING
AND DISAPPEARING
"BRIDGES"
BETWEEN
SINGULARITIES

PERHAPS
MATTER
ENTERS
THE UNIVERSE
FROM
THE
SINGULARITY

HAVING TRAVELED
THROUGH
SPACE-TIME
"BRIDGES"
FROM
ANOTHER
PLACE

AS IT PASSES
THE FINAL
OUTER
EVENT HORIZON
OF THE
WHITEHOLE

IT
FLASHES
INTO BEING
IN WHAT
APPEARS
TO BE

THE
CREATION
OF
LIGHT!

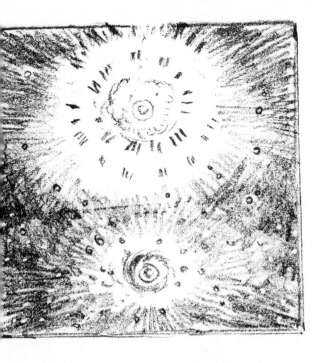

PERHAPS A BLACKHOLE
IN THIS UNIVERSE
IS A WHITEHOLE
IN ANOTHER
UNIVERSE..

OR IN ANOTHER
LOCATION
IN THIS
UNIVERSE.

OR, PERHAPS, ONE IN THE SAME,
WHERE THE PASSING OF THE OLD
IS THE BIRTH OF THE NEW!

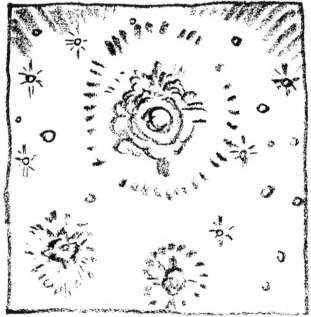

PERHAPS
QUASARS
AND
BLAZARS

AND
THE GALAXIES
THEMSELVES

ARE DIFFERENT
PHASES OF
THE SINGULARITY

LIGHTING THE SKIES IN SOME GREAT
COSMIC LIGHT SHOW!

WHITEHOLES

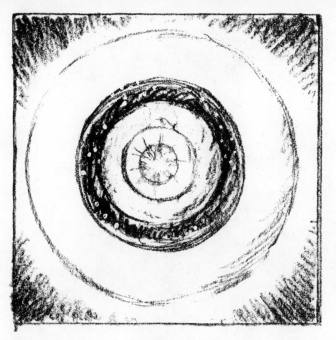

A WHITEHOLE SINGULARITY   ACTS BEYOND THE KNOWN LAWS OF SCIENCE

THERE ARE BRILLIANT GALAXIES NEAR THE CENTER..   IS IT THE QUALITY OF LIGHT ?   · OR ·   ARE THEY MOVING AT INCREDIBLE SPEEDS ?

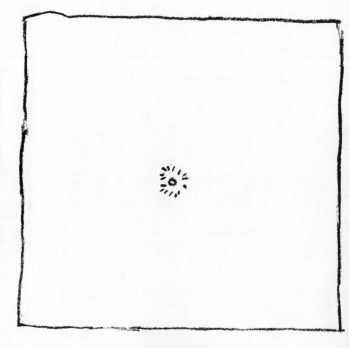

WHITEHOLES COULD RANGE IN SIZE FROM THOSE THAT MIGHT BE AT THE CENTER OF A GALAXY   ..BILLIONS OF TIMES MORE MASSIVE THAN THE SUN...   WAY DOWN TO THOSE AS TINY AS THE NUCLEUS OF AN ATOM

IN RECENT YEARS THERE HAVE BEEN SOME HIGHLY IMAGINATIVE COMBINATIONS OF THE QUANTUM THEORY AND RELATIVITY   ACCORDING TO SOME THEORIES, LOOKING AT A ROTATING BLACKHOLE (WHITEHOLE?) - LIGHTHOLE ·   WE SHOULD BE ABLE TO SEE THE EARTH AT ANY TIME IN "THE PROBABLE PAST"...   ...AND AT ANY TIME IN THE "PROBABLE FUTURE"!

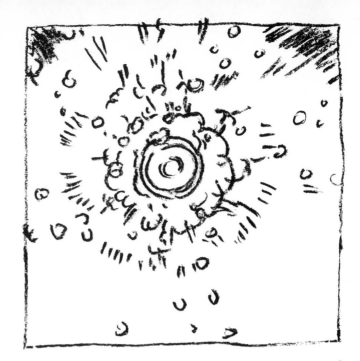

HITEHOLE
INGULARITY
T THE CENTER
F
HE GALAXY

MIGHT
CREATE
ENERGY AND
LIGHT
FROM
A KIND OF
"NOTHINGNESS"

IF THERE WAS
A WHITEHOLE
SINGULARITY
AT
THE CENTER
OF
THE GALAXY

NEW LIGHT
AND
MATTER
MIGHT
FLASH
FORTH

IN THE WAKE
OF THE
OLD STARS
THAT
DISAPPEAR
WITHIN

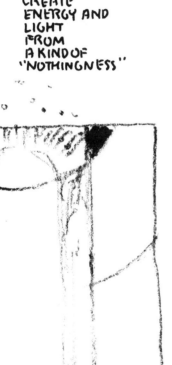

OME THEORIES TELL OF PASSAGES
HROUGH THE SINGULARITY

INTO
OTHER
UNIVERSES

AND

INTO OTHER
TIMES AND
PLACES

THE GREAT LIGHTWAY?

AT THE
EVENT
HORIZON..
ALL TIME
COMES TO
A STOP!!

A MODERATELY ROTATING
"LIGHTHOLE"
HAS TWO EVENT HORIZONS
AROUND A
"RING SINGULARITY"

ACCORDING TO SOME,
IF "LIGHTHOLES"
EXIST..
WE COULD TRAVEL
IN AND OUT OF
THE EVENT HORIZONS

THROUGHOUT TIME..
INTO THE PASTS
AND FUTURES..

AND EVEN TRAVEL
THROUGH THE
SINGULARITY!
INTO
IMAGINARY
UNIVERSES!

A LOOK AT THE GALAXIES
MOVING IN THE SKIES
OVER THOUSANDS OF
THOUSANDS OF YEARS

REVEALS A MOST WONDERFUL
AND GRACEFUL
SYMPHONY OF
LIGHT

AS STARS COME AND GO
AND THOUSANDS OF
THOUSANDS OF GALAXIES
LIGHT AND DIM

ASTRONOMERS HOPE ONE DAY
TO BE ABLE TO SEE
THE BEGINNINGS OF
GALAXIES

PERHAPS THEY WILL THEN KNOW
MORE ABOUT HOW OUR
OWN GALAXY
CAME TO BE

OLD STORIES

TALES OF TALES

FOR THOUSANDS OF YEARS
BEFORE THE BEGINNING OF
THE WRITING OF HISTORY

IN THE EVENING
FRIENDS AND FAMILY
GATHERED TOGETHER
TO REST
TO SLEEP UNDER THE STARS

INSPIRED BY THE STARS

SOME TOLD STORIES
OF WHAT THEY SAW

SOME TOLD THEIR CHILDREN . .
WHO TOLD THEIR CHILDREN . . .
WHO TOLD THEIR CHILDREN . . .

SOME TELL OF AN
ETERNAL GOD OR
GODS AND GODESSES
CREATING
THE WORLD

IN MANY MYTHOLOGIES
THE FORCES OF NATURE
WERE PERSONALIZED
AS "GODS"

IN A NORSE MYTH,
OUT OF "NOTHING"
IN A GAPING CHASM
TWO GODS
MADE THE WORLD
AT THE CENTER,
WATERED BY
THREE
ETERNAL SPRINGS
APPEARED
THE TREE OF LIFE

IN A BABYLONIAN MYTH,
A GREAT GOD
"ORGANIZED" THE COSMOS
BY DEFEATING THE
SEA OF PRIMITIVE
CHAOS . .
FROM WHICH
AROSE THE
WORLD

IN A MESOPOTAMIAN MYTH
THE GODS WERE OF
THE FORCES OF NATURE
OF THE EARTH
. . AND THE BEGINNING
OF THE WORLD
WAS THE BALANCING
OF THE FORCES.

SOME TELL OF
A KIND OF
"NOTHINGNESS"
FROM WHICH
HEAVEN AND EARTH
AND THE GODS
THEMSELVES
WERE CREATED

IN A POLYNESIAN MYTH,
OUT OF A CHAOTIC VOID,
THERE FIRST
EVOLVED MOVEMENT.
THEN SOUND . . THEN
A PULSATING LIGHT . .
THEN HEAT . . THEN
MOISTURE . . THEN
MATTER . . THEN FORM . .
THEN FATHER HEAVEN
AND MOTHER EARTH . .
WHO THEN CREATED ALL
GODS AND ALL EARTHLY
BEINGS AND ALL
THINGS OF NATURE

IN AN ICELANDIC MYTH,
OUT OF A "YAWNING
ABYSS" . .
IN THE LAND OF THE
NORTH . . OUT OF THE
CLOUDS AND SHADOWS . .
A FOUNTAIN APPEARED
FROM WHICH SPRANG
TWELVE RIVERS OF
ICE . .
AND IN THE LAND OF
THE SOUTH . . CAME
OTHER RIVERS . .
WHERE THEY MET . .
THE ICE MELTED
INTO DROPS . .
FROM WHICH THE
FIRST LIVING BEING
WAS FORMED

IN A GREEK MYTH
FROM A
VAST DARK CHAOS"
THERE APPEARED
THE EARTH . .
THEN LOVE APPEARED
TO WATCH OVER
THE FORMATION
OF EVERYTHING . .
THE DAY AND NIGHT
THE SKY AND STARS
AND THE SEAS.

IN A ROMAN MYTH . .
. . BEGINNING WITH A
FORMLESS CONFUSION

THE GOD OF GODS
SEPARATED
AIR, FIRE, EARTH
AND WATER . .
AND THEN
BECAME
THE GOD OF
BEGINNINGS . .
OF THE SUN

IN AN EGYPTIAN MYTH
THE SPIRIT OF ALL
EXISTENCE . .
IN A PRIMORDIAL
OCEAN . .
WEARY OF IMPERSONAL
EXISTENCE . .
ENCLOSED HIMSELF
IN A LOTUS BUD
TO RISE FROM
THE "ABYSS" . . .
THEN FORMED
THE LAND . .
AND CREATED
ALL THE GODS "

IN A HINDU MYTH,
THE GREAT GOD
CREATED THE WATERS
INTO WHICH HE
PLACED A SEED . .
WHICH BECAME
A HEAVENLY "EGG"
WHICH HE
SPLIT APART . .
THE HEAVENS
SPRINGING FORTH
FROM THE
GOLDEN HALF . .
AND THE EARTH
FROM THE
SILVER HALF

IN AN OCEANIC MYTH
A MOTIONLESS, QUIET
"BEING" CAME TO BE
OUT OF A MIST . .
GIVING BIRTH
TO ANOTHER . . WHO
PASSED ON . . AND
FROM WHOSE HEART
A TREE GREW . .
AND FROM THE BUDS
OF THE TREE . .
ALL THE GODS AND
ALL THE EARTHLY
BEINGS WERE BORN

IN A HAWAIIN MYTH,
ALL THINGS CAME FROM
A "SHADOWY VOID"
WHICH CAME FROM
THE WRECKAGE OF
A PREVIOUS "WORLD"

IN AN ASSYRO-BABYLONIAN
MYTH . THE MELTING OF
TWO KINDS OF WATERS . .
"SWEET" AND "SALT" WATERS
(THE ORGANIZING FORCE
AND THE TUMULTUOUS
FORCE) . .
WAS THE BEGINNING
OF THE GODS
AND THE EARTH
AND THE SKIES

TO TRULY UNDERSTAND
THE ESSENCE OF THESE
CREATION STORIES,
ONE WOULD HAVE TO
UNDERSTAND THE
LANGUAGE IN WHICH
THE ORIGINALS WERE
SPOKEN OR WRITTEN

ONE WOULD HAVE TO
UNDERSTAND THE
"IMAGES" OF THE TIMES
AND THE NATURE
AND CHARACTER OF
THE LIFE OF THE TIMES

THESE LITTLE STORIES
ARE ENGLISH LANGUAGE
VERSIONS OF "VERSIONS
OF VERSIONS" OF
THE ORIGNAL STORIES . .

WHAT DO YOU
SEE IN THE
STARS?

WHAT DO YOU
SEE IN THE
SKIES
?

THE ETERNITY
THE LIGHT OF THE STARS
ROUGHOUT TIME

W STARS APPEAR
O STARS DISAPPEAR

LOOKING TO
THE 'STARS
ONE
WONDERS..

IF THERE WAS
A BEGINNING
TO ALL OF
THIS...

WHAT WAS
HERE
BEFORE
THE
"BEGINNING"?

SOME TELL OF
AN ETERNAL WORLD
WHEREIN
HEAVEN
AND EARTH
HAVE EXISTED
FOREVER

IN SOME
MYTHS,
THE BALANCING
OF THE
"FORCES OF
NATURE"
WAS THE
ESSENCE
OF THE
"BEGINNING"

IN SOME MYTHS,
THE CREATION OF
"ORDER"
OUT OF CHAOS
WAS THE
ESSENCE
OF THE
"BEGINNING"

A GREEK MYTH,
RIVER "OCEANUS"
S THE FATHER
ALL THINGS..
SKY
O THE EARTH
VE BIRTH
THE GODS"

A 4TH CENTURY B.C.
PHILOSOPHER
TOLD OF AN
INDESTRUCTIBLE,
"ETERNAL" WORLD
WITHOUT A
BEGINNING

SOME TELL OF
AN ETERNITY
WHEREIN
SOME UNIVERSAL
PROCESS
CREATES
THE WORLD

IN A PHOENICIAN
MYTH..
IN "COSMIC TIME"
WHICH CONTAINS
ALL THINGS..
CAME DARKNESS
AND DESIRE..
WHICH PRODUCED
INTELLIGENCE..
AND THE MOVEMENT
OF LIFE..
WHICH PRODUCED
"THE COSMIC
EGG"

IN THE VIEW OF
A 6TH CENTURY B.C.
PHILOSOPHER..
THERE IS AN
UNLIMITED "INFINITE
AND ETERNAL
MOTION"
CAUSING OPPOSING
PROPERTIES
TO SEPARATE OUT
TO MAKE
THE WORLD

IN A 5TH CENTURY B.C.
PHILOSOPHY..
THE "UNIVERSAL MIND"
STARTED THE COSMOS
REVOLVING..
REVOLVING IN LARGER
AND LARGER REGIONS..
OVER THE YEARS..
AND SEPARATING OUT
ALL THINGS

A 6TH CENTURY B.C.
PHILOSOPHER
PRESENTED
"DIVINE PRINCIPLES"
AS BEING THE
FOUNDATION OF
ALL CREATION..
THE WORLD HAVING
BEEN CREATED
IN ACCORDANCE
WITH THESE
PRINCIPLES

SOME TELL OF
AN ETERNAL HEAVEN
FROM WHICH
THE EARTH
WAS CREATED

IN A MYTH OF THE
PAWNEE INDIANS OF
THE PLAINS OF
NORTH AMERICA..

FROM THE HEAVENS
WHERE THE GREAT CHIEF
AND THE GODS OF THE
SUN AND STARS LIVED..

IN CLOUDS OF THUNDER
AND LIGHTNING..

THE GREAT CHIEF
DROPPED A PEBBLE
AND THE CLOUDS
BECAME A SEA..
AND THE FOUR STARS
OF THE FOUR CORNERS
OF HEAVEN
SEPARATED THE WATERS..
AND THE LAND
APPEARED

SOME TELL OF
AN ETERNAL GOD
:ONE ALONE:
CREATING
HEAVEN
AND EARTH

OLD TESTAMENT..
E BEGINNING
REATED THE HEAVEN
HE EARTH"

SEPARATING THE LIGHT
THE DARKNESS.

-AS HIS SPIRIT HOVERED
THE WATERS..

D "LET THERE BE LIGHT"
THERE WAS LIGHT

IN THE SECOND DAY
HE MADE A "FIRMAMENT"
DIVIDING THE WATERS..
UNDER THE FIRMAMENT
FROM THE WATERS
ABOVE THE FIRMAMENT
WHICH HE CALLED
HEAVEN

IN THE THIRD DAY
HE GATHERED THE
WATERS UNDER HEAVEN
INTO ONE PLACE..
CREATING THE SEAS
AND THE LAND
OF EARTH..
AND THE LAND
BROUGHT FORTH
GRASS AND SEEDS
AND FRUIT TREES..

IN THE FOURTH DAY
HE MADE A GREAT LIGHT
FOR THE DAY
AND A LESSER LIGHT
FOR THE NIGHT

AND PUT THEM IN THE
FIRMAMENT OF HEAVEN
TO DIVIDE THE LIGHT
FROM THE DARKNESS

IN THE FIFTH DAY
HE CREATED
BIRDS IN THE SKY
AND FISH AND
LIVING THINGS
OF THE WATERS

IN THE SIXTH DAY
HE CREATED
THE ANIMALS
OF THE EARTH

AND HE CREATED,
IN HIS OWN IMAGE,
MAN AND WOMAN

ON THE
SEVENTH DAY
GOD RESTED
AND BLESSED
AND HALLOWED
THIS DAY

THROUGHOUT HISTORY
THERE HAVE BEEN
THOUSANDS OF STORIES
OF "THE BEGINNING"

SOME WERE
DIVINELY
INSPIRED

MANY WERE
VERSIONS OF VERSIONS
OF POPULAR STORIES
OF OTHER
CIVILIZATIONS
OR TIMES

MANY ANCIENT MYTHOLOGIES
TELL OF AN ETERNAL WORLD . .

A WORLD WHERE
HEAVEN AND EARTH AND THE SEAS
HAVE ALWAYS EXISTED.

OTHERS
TELL OF HOW
HEAVEN AND EARTH AND THE SEAS
CAME TO BE

FOR SOME . .
THE WORLD WITH ITS GODS
ORIGINATES OUT OF
" A KIND OF NOTHINGNESS"

OTHERS
TELL OF AN
ETERNAL HEAVEN
FROM WHICH
THE EARTH AND SEAS
CAME TO BE

FOR SOME..
AN ETERNAL GOD..
ALONE
CREATES THE HEAVENS
AND THE EARTH

THROUGHOUT HISTORY
IDEAS APPEAR
AND DISAPPEAR

VERSIONS OF VERSIONS
OF IDEAS
COME AND GO

TAKING ON THE
"COLOR"
OF THE TIMES

THERE ARE TALES OF
THE BEGINNING OF THE WORLD

COMING OUT OF
THE BALANCING OF
THE FORCES OF NATURE

THERE ARE TALES
OF GODS AND WORLDS

APPEARING OUT
OF A VAST, DARK
CHAOS...

THERE ARE TALES OF
GODS AND WORLDS
APPEARING OUT OF
"NOTHINGNESS"

THERE ARE TALES
OF ALL THINGS
BEGINNING OUT OF
A QUIET "MIST"

THERE ARE
TALES OF AN
ETERNAL
WORLD

THERE ARE TALES OF
HEAVENS AND EARTHS
ARISING
FROM WITHIN
GREAT OCEANS AND RIVERS

THERE ARE TALES
OF NEW WORLDS
ARISING FROM THE
REMAINS OF THE OLD

THERE ARE TALES
OF AN
ETERNAL "COSMIC TIME"
WHICH CONTAINS
ALL THINGS

THERE ARE
TALES OF
THE WORLD
BEGINNING
WITHIN
A
GREAT
EVERPRESENT
REALITY

THERE ARE TALES
OF
HEAVENS APPEARING
ON EARTH

THROUGHOUT HISTORY,
WHILE SOME HAVE
WONDERED
WHERE
"ALL THIS"
COMES FROM

OTHERS HAVE
WONDERED
ABOUT HOW IT
WORKS
AND WHAT IS
THE
"BIG PICTURE"

HERE IS A QUICK TOUR
THROUGH TIME,
HIGHLIGHTING SOME OF
THE DRAMATIC CHANGES
IN THE "PICTURE"
OF THE UNIVERSE

THE CELESTIAL
VAULT IS NOW
A DOME

ABOUT 5,000 YEARS AGO,
THE WORLD WAS FLAT,
AND THE STARS WERE
ATTACHED TO A
"CELESTIAL VAULT"
ABOVE

ABOUT 4,000 YEARS AGO
THE WORLD WAS A
CIRCULAR DISC
SURROUNDED BY
WATER

HUNDREDS
OF THOUSANDS
OF STARS

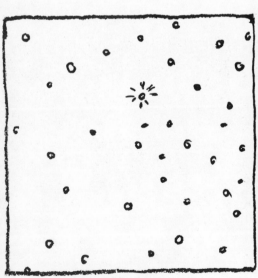

ABOUT 370 YEARS AGO,
50,000 STARS
CAME TO LIGHT
THROUGH THE
FIRST ASTRONOMICAL
TELESCOPE

ABOUT 200 YEARS AGO,
THE SUN "BEGAN"
TO MOVE IN SPACE..
AS MORE AND MORE
STARS AND LIGHTS
APPEARED

ABOUT 100 YEARS AGO,
HUNDREDS OF THOUSANDS
OF STARS
BEGAN TO
APPEAR

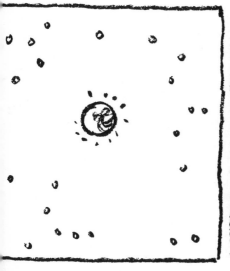

ABOUT 2,500 YEARS AGO,
THE EARTH WAS,
FOR A BRIEF "MOMENT",
THOUGHT BY SOME
TO BE A
SPHERE

ABOUT 2,150 YEARS AGO
THE FLAT EARTH
CAME TO BE
THE CENTER
OF A UNIVERSE
OF GREAT REVOLVING
TRANSPARENT
SPHERES.

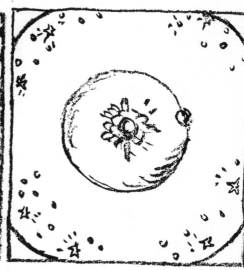

ABOUT 400 YEARS AGO,
THE SUN BECAME
THE CENTER
OF THE UNIVERSE . .
AND THE EARTH
WAS NOW REVOLVING
ABOUT THE SUN.

ABOUT 50 YEARS AGO,
THOUSANDS OF GALAXIES
APPEARED
BEYOND
THIS
GALAXY

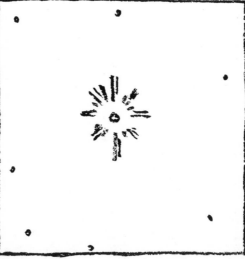

ABOUT 20 YEARS AGO,
THE FIRST
QUASARS
CAME TO
LIGHT

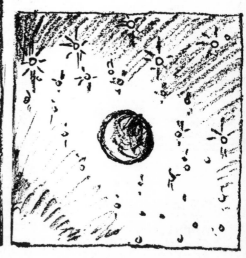

IN ANCIENT CIVILIZATIONS
AS, PERHAPS, IT WAS
IN TIMES BEFORE THE
WRITING OF HISTORY..

ASTRONOMY WAS
INTERWOVEN WITH
THE MYTHOLOGIES OF
THE TIMES

 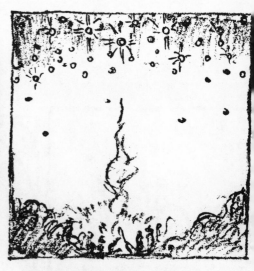

GODS AND SPIRITS AND
MYTHOLOGICAL "BEINGS"
WERE THOUGHT TO LIVE
IN THE SKIES..
CELESTIAL "EVENTS"
WERE THOUGHT TO BE
THEIR ACTIONS

THE EARLIEST CIVILIZATIONS
MAY HAVE WATCHED THE
POSITIONS OF THE SUNS
RISING AND SETTING
AND THE MOVEMENTS OF
THE MOON AND THE STARS
TO DETERMINE DIRECTIONS
AND.. LATER.. TO DETERMINE
THE BEGINNINGS OF THE SEASONS

THE CONSTELLATIONS
OF THE ZODIAC
APPEAR TO MOVE
ALONG THE SAME
PATH

 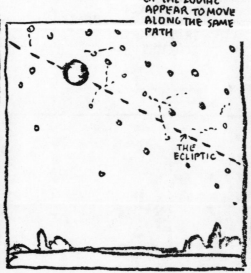 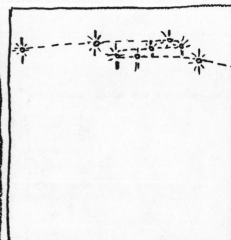

THE
ECLIPTIC

THE ASTRONOMER /
ASTROLOGERS OF
BABYLONIA WERE
AMONG THE FIRST TO
CAREFULLY RECORD
THE MOVEMENTS OF
THE STARS..

THEY MADE THE FIRST
KNOWN "STAR-CATALOG"
AROUND 1800 B.C.

THEY RECORDED
CELESTIAL "EVENTS"
CONTINUOUSLY FROM
ABOUT 747 B.C.
(POSSIBLY EARLIER)
AND THEY RECORDED
AND COULD PREDICT
THE MOVEMENTS OF
THE PLANETS..
KNOWN AS THE
"WANDERER" STARS

THE ASTRONOMERS
OF THE EARLY
CIVILIZATIONS
OF THE MESOPOTAMIAN
VALLEY.. AND
THE ASTRONOMERS
OF ANCIENT CHINA
KNEW OF THE
"CELESTIAL EQUATOR"
AND OF THE "ECLIPTIC".
THE PATH OF THE SUN..
AND THE WANDERERS

THE "WANDERERS"
WOULD TRAVEL THE
SAME GENERAL PATH
IN THE SKY..

AT TIMES SLOWING
DOWN TO STOP..
REVERSE DIRECTION..
THEN STOP AGAIN..
AND RESUME THEIR
ORIGINAL DIRECTION

ERE IS SOME EVIDENCE AT THE POSITIONS OF E SUN AND THE MOON ERE OBSERVED AS AR BACK AS IN THE OLITHIC ERA!

THE SITE OF STONEHENGE IN ENGLAND MIGHT HAVE BEEN IN USE IN ABOUT 2700 B.C... BEFORE THE GREAT STONES WERE PUT IN PLACE ABOUT 2,000 B.C.

PERHAPS, TO NOTE THE BEGINNINGS OF THE SEASONS · · THE LONGEST AND SHORTEST DAYS OF THE YEAR · · AND MORE.

THERE ARE MANY OTHER EARLY STONE "OBSERVING" SITES.. LITTLER THAN, BUT SIMILAR TO STONEHENGE

THE "MEDICINE WHEEL" AT THE BIGHORN MOUNTAINS IN THE U.S.A. IS A STONE "OBSERVATORY" OF LATER TIMES

THE APPARENT RETROGRADE MOTION OF A PLANET BEYOND THE EARTHS ORBIT

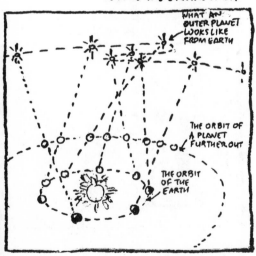

WHAT AN OUTER PLANET LOOKS LIKE FROM EARTH

THE ORBIT OF A PLANET FURTHER OUT

THE ORBIT OF THE EARTH

XPLANATIONS ERE ATTEMPTED · · BUT IT WAS NOT NTIL MANY CENTURIES ATER THAT THESE WANDERER STARS WERE UNDERSTOOD TO BE THE PLANETS OF THE SOLAR SYSTEM ORBITING THE SUN.

AS THE EARTH ORBITS THE SUN QUICKER THAN THE PLANETS FURTHER FROM THE SUN · · THOSE PLANETS WOULD "APPEAR" TO BE REVERSING DIRECTION AT TIMES. AS VIEWED FROM EARTH

THE EARLY ASTRONOMERS HAD NOT MUCH MORE THAN SIMPLE WOODEN CALCULATING AND SIGHTING DEVICES.

AS THERE WERE NO TELESCOPES UNTIL THE 17TH CENTURY

THROUGHOUT MOST OF HISTORY · · ALL OF THE THEORIES ABOUT THE STRUCTURE OF THE UNIVERSE WERE BASED UPON WHAT COULD BE SEEN AT THE TIME · · ABOUT A THOUSAND STARS!

LETS LOOK AGAIN .. BEGINNING AT THE BEGINNINGS OF THE WRITING OF HISTORY

THE VAULTED SKY

THE PICTURE OF "THE WORLD" HAS CHANGED DRAMATICALLY THROUGHOUT HISTORY

IN ANCIENT TIMES, MOST PEOPLE THOUGHT THE WORLD WAS THE WHOLE UNIVERSE

IN ANCIENT TIMES, THE EARTH WAS THOUGHT TO BE FLAT .. COVERED BY A VAULTED SKY .. A "FIRMAMENT" .. IN WHICH THE STARS WERE EMBEDDED

IN THE EARLY YEARS OF EARLY GREEK TIMES, THE EARTH WAS A CIRCULAR DISC SURROUNDED BY THE RIVER "OCEANUS"

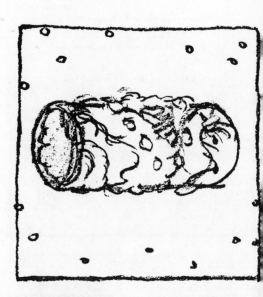

ABOUT 550 B.C .. THE PHILOSOPHER ANAXIMANDER SUGGESTED THE EARTH WAS CYLINDRICAL ..
TO EXPLAIN THE SHIPS DISAPPEARING OVER THE HORIZON.

IN BABYLONIAN TIMES,
THE EARTH WAS A FLAT DISC
SURROUNDED BY A SEA..
AND THE SKY WAS A
SOLID "FIRMAMENT"
VAULTED ABOVE IT ALL

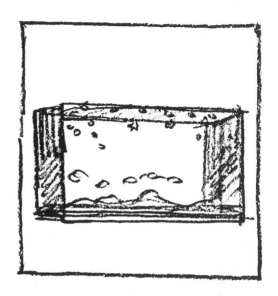

IN EGYPTIAN TIMES,
THE EARTH WAS A
RECTANGLE, AND
ALL THE UNIVERSE
WAS IN THE SHAPE
OF A RECTANGULAR
BOX

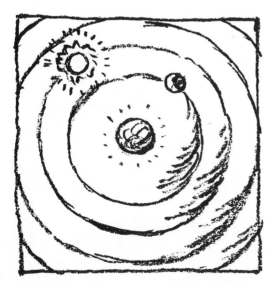

THE PHILOSOPHER
PYTHAGORUS
BELIEVED THAT
"THE MOST PERFECT"
OBJECT WAS

A SPHERE
MOVING
UNIFORMLY

ABOUT 530(?) B.C.
HE PRESENTED A
PICTURE OF THE
WORLD AS A <u>SPHERE</u>

..AROUND WHICH
THE WANDERER STARS
WERE MOVING ON
"HEAVENLY SPHERES"

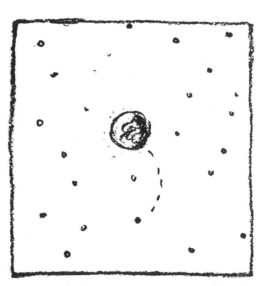

ABOUT 450 B.C.
THE PHILOSOPHER
PHILOLAUS
DESCRIBED
THE WORLD
AS A SPHERE .
<u>MOVING THROUGH
SPACE !!</u>

THIS IDEA
DIDN'T
LAST
VERY
LONG

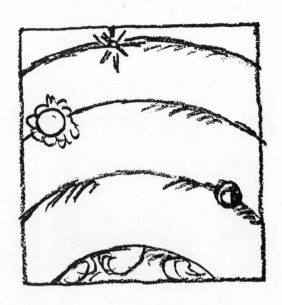

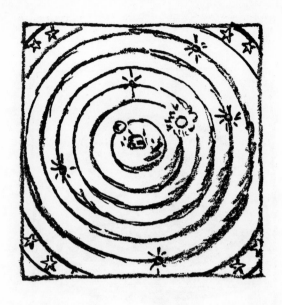

THE IDEA OF CONCENTRIC SPHERES AROUND THE EARTH WAS DEVELOPED BY PLATO

IN THE 4TH CENTURY, B.C. EUDOXUS IMAGINED TRANSPARENT SPHERES REVOLVING ABOUT THE EARTH.. MOVING SLOWLY TOGETHER FROM EAST TO WEST.

THE THOUSANDS OF "FIXED" STARS WERE ATTACHED TO THE OUTERMOST SPHERE..

AND THE 7 "WANDERERS" (THE FIVE STAR PLANETS, THE SUN AND THE MOON) WERE ATTACHED TO 7 CONSECUTIVELY LITTLER SPHERES WITHIN SPHERES.

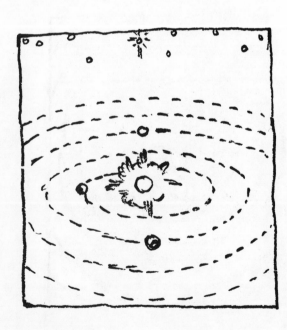

IN THE 3RD CENTURY B.C. THE ASTRONOMER ARISTARCHUS PRESENTED A "HELIOCENTRIC" SUN-CENTERED UNIVERSE

THE EARTH BEING ONE OF SEVERAL PLANETS REVOLVING ABOUT THE SUN.. AND THE STARS BEING VERY FAR AWAY

...VERY CLOSE TO THE PICTURE OF THE SUN AND THE PLANETS WE HAVE TODAY!

IN THE 2ND CENTURY, B.C. ERATOSHENES CALCULATED THE EARTH'S DIAMETER TO BE 8,000 MILES ACROSS.. VERY CLOSE TO WHAT IT IS!

BY ABOUT 150 B.C. THE EARLY ASTRONOMERS HAD PRETTY WELL FIGURED OUT THE SIZE AND SHAPE OF THE EARTH.. AND THE DISTANCE OF THE MOON FROM EARTH

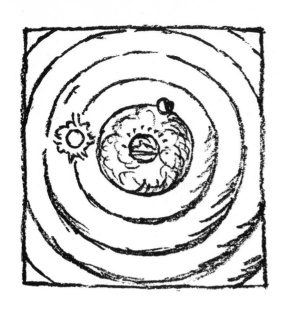

ARISTOTLE
PRESENTED A
"HEAVENLY SPHERE"
UNIVERSE OF
CONCENTRIC
SPHERES...
TO EXPLAIN THE
RETROGRADE MOTION
OF THE WANDERERS

THE EARTH AT THE CENTER
AND EVERYTHING UP TO
AND WITHIN THE
SPHERE OF THE MOON
WAS MADE OF:
EARTH,
AIR,
WATER,
FIRE

IN THE 4TH CENTURY B.C.
HERACLIDES PRESENTED
THE PLANETS (WANDERERS)
REVOLVING ABOUT
THE SUN.

WHILE THE SUN
REVOLVED ABOUT
A ROTATING EARTH...
FIXED AT THE CENTER
OF THE UNIVERSE

ABOUT 134 B.C.,
THE ASTRONOMER
HIPPARCHUS
PRESENTED THE
"FIRST GREAT
STAR MAP"
OF ABOUT
1000 STARS

AND CLASSIFIED
ALL THE STARS
KNOWN AT THE
TIME INTO
6 MAGNITUDES
OF
BRIGHTNESS

THE BRIGHTEST
TO BE CALLED
"FIRST
MAGNITUDE
STARS"...
A SYSTEM
STILL IN
USE TODAY

MORE THAN 250 YEARS
LATER, AROUND 130 A.D.
THE ASTRONOMER
PTOLEMY
INCREASED THE
STAR MAP TO MORE THAN
1,000 STARS! (1,028)

AND DESIGNATED
48 CONSTELLATIONS
WHEREIN THEY ALL
WERE LOCATED

PTOLEMY DEVELOPED
THE CRYSTAL SPHERE IDEA
TO EXPLAIN
THE MOTION OF
THE WANDERER
STARS

THE CRYSTAL SPHERES

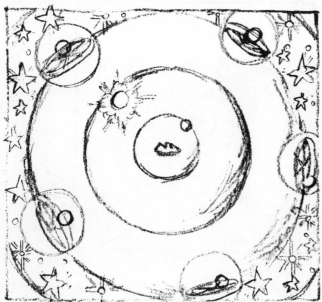

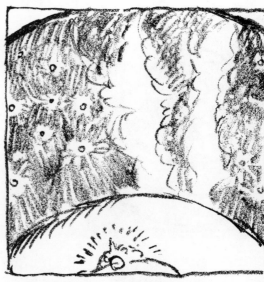

THE MILKY W
WAS SEEN A
A REGION OF
THE HEAVEN
"AS WHITE AS
MILK"

VARYING IN
WIDTH AND
BRIGHTNESS
WITH A VARI
OF ARRAYS
OF STARS

THE "PTOLEMAIC" UNIVERSE OF THE CRYSTAL SPHERES BECAME FIRMLY ESTABLISHED IN THE 2ND CENTURY A.D.

FOR ABOUT 14 CENTURIES THEREAFTER..

IT WAS GENERALLY ACCEPTED THAT THE SUN.. AND ALL THE STARS: WERE REVOLVING ABOUT THE EARTH - ...

THE EARTH AT THE CENTER OF A NUMBER OF GREAT REVOLVING CRYSTAL SPHERES

ALL THE STARS WERE RIDING THE GREATEST REVOLVING CRYSTAL SPHERE

THE SUN AND THE PLANETS WERE ATTACHED TO ANOTHER GREAT CRYSTAL SPHERE.. CLOSER TO EARTH

AND THE MOON WAS ATTACHED TO THE LITTLES CLOSEST CRYSTAL SPHERE

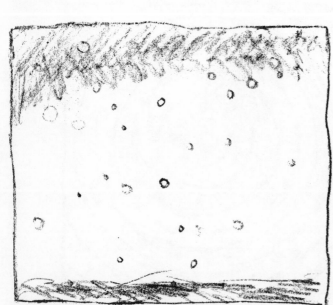

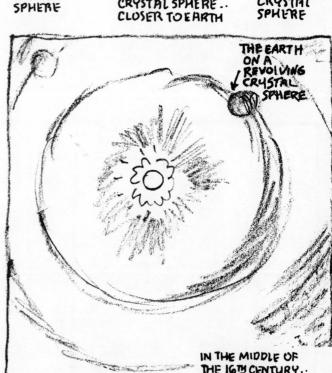

THE EARTH ON A REVOLVING CRYSTAL SPHERE

WITH THE FALL OF THE ROMAN EMPIRE, THE ARABIAN CIVILIZATIONS BECAME HEIR AND PRESERVER OF THE KNOWLEDGE OF THE GREEKS.. AND OF THE KNOWLEDGE AND MAPS FROM INDIA AND FROM CHINA

IN THE 9TH CENTURY, THE "HOUSE OF WISDOM" WAS ESTABLISHED IN BAGHDAD TO PRESERVE THE WORLD'S ASTRONOMICAL MAPS AND WRITINGS .. AND TO CONTINUE THE SIGHTING AND MAPPING OF THE SKIES AND THE DISCOVERY OF THE STARS.

THE WORLD PASSED THROUGH THE SHADOW OF THE DARK AGES

BY THE 15TH CENTURY, THE IDEA OF AN INFINITE UNIVERSE HAD REAPPEARED ..

A UNIVERSE WITHOUT LIMIT TO WHAT COULD BE KNOWN.

.. OF STARS LIGHTING SPACE THAT GOES ON FOREVER ..

_ OF A UNIVERSE WITHOUT A CENTER .. WITHOUT EDGE .

IN THE MIDDLE OF THE 16TH CENTURY.. THE IDEA OF A SUN-CENTERED UNIVERSE REAPPEARED..

THE CHURCHMAN/SCHOLA NICOLAUS COPERNICUS BROUGHT TO LIGHT A REVOLUTIONARY IDE OF THE TIME ..

THAT THE SUN WAS AT THE CENTER.
AND THE EARTH WAS REVOLVING ABOUT THE SUN!

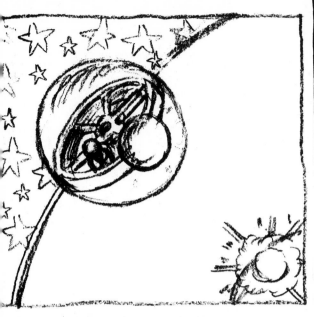

WITHIN CRYSTAL SPHERES THAT WERE ATTACHED TO THE MIDDLE SPHERE..

THE 5 KNOWN PLANETS (WANDERER STARS) RODE ON WHEELS WITH WHEELS WITHIN THE WHEELS!

ALTHOUGH ARISTARCHUS HAD PICTURED A SUN-CENTERED UNIVERSE ABOUT 500 YEARS EARLIER.. AND ALTHOUGH MANY PHILOSOPHERS AND ASTRONOMERS THROUGHOUT HISTORY HAD IMAGINED THE EARTH AS A SPHERE..

THE IDEA OF AN EARTH-CENTERED UNIVERSE ..

THE SUN AND THE STARS REVOLVING ABOUT A FLAT EARTH..

WAS THE MOST POPULAR VIEW OF THE UNIVERSE FOR ABOUT 1700 YEARS!

IN THE LATE 16TH CENTURY THE ASTRONOMER TYCHO BRAE CAREFULLY WATCHED THE SKIES FOR 20 YEARS.. MEASURING THE MOVEMENTS OF 777 STARS AND THE 5 KNOWN PLANETS..

DOING IT ALL WITHOUT A TELESCOPE!!

HE PICTURED A "UNIVERSE" SIMILAR TO THAT OF HERACLIDES.. WITH THE SUN REVOLVING ABOUT THE EARTH.. AND THE PLANETS REVOLVING ABOUT THE SUN

..BUT..A UNIVERSE WITHOUT THE CRYSTAL SPHERES (AFTER OBSERVING THE GREAT COMET OF 1577-GO RIGHT THROUGH THEM.. HE REASONED THEY COULDN'T BE THERE

IN THE FIRST YEAR OF THE 17TH CENTURY, THE ASTRONOMER J. KEPLER CAME TO WORK WITH TYCHO .. WHO PASSED ON A YEAR LATER ..

.. AND KEPLER CONTINUED THE MAPPING OF THE HEAVENS

50,000 STARS

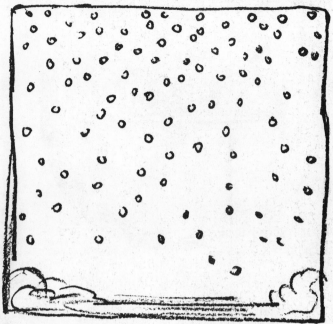

THROUGHOUT HISTORY,
ABOUT 2,000 STARS
COULD BE SEEN IN THE
HEAVENS ALL OVER
THE WORLD..
THOUGHT TO BE ALL
OF THE STARS OF
ALL OF CREATION.

IN THE BEGINNING
OF THE 17TH CENTURY..
IN THE YEAR 1609..
THE ASTRONOMER
GALILEO GALELI
GAZED THROUGH
THE FIRST
ASTRONOMICAL
TELESCOPE !

ALL AT ONCE,
THE SKIES WERE FILLED
WITH
50,000 STARS !

THE HAZY "MILKY WAY"
WAS SEEN TO BE
MADE OF
THOUSANDS OF
STARS !

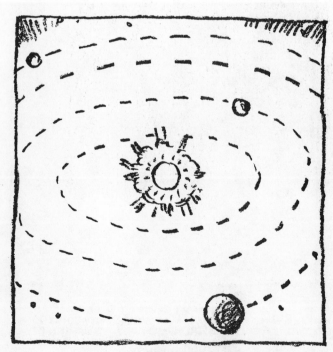

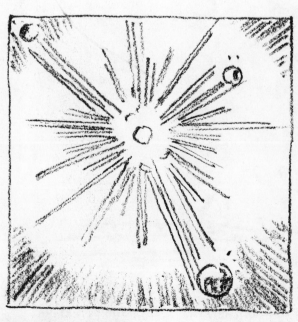

THE ASTRONOMER KEPLER,
NOW ALSO USING A TELESCOPE,
DETERMINED ALL THE PLANETS
WERE INDEED ORBITING
THE SUN.. THOSE PLANETS
CLOSEST TO THE SUN
MOVING THE FASTEST..
ALL IN ELLIPTICAL ORBITS

NOT KNOWING ABOUT
GRAVITY. HE IMAGINED
IT WAS THE SUN'S RAYS
THAT RODE THE
PLANETS AROUND
IN THEIR ORBITS

MOUNTAINS
ON
THE
MOON

THE
MOONS
OF
JUPITER

THE
SHADOWS
OF
VENUS

THE SUN
AT
THE CENTER

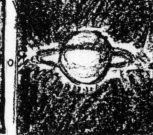

FOR THE
FIRST TIME,
THE SUN
WAS
SEEN
TO BE
ROTATING

FOR THE
FIRST TIME,
MOUNTAINS
COULD BE
SEEN ON
THE MOON

FOR THE
FIRST TIME,
MOONS WERE
SEEN
REVOLVING
ABOUT
JUPITER

AND THE PLANET
VENUS, NOW SEEN
UP CLOSE . .
TRULY SEEMED
TO BE REVOLVING
AROUND THE
SUN . .

AS ITS SHADOWS
GOT LITTLER AND
THEN GOT BIGGER
OVER TIME

YEARS LATER.
WHAT GALILEO
SAW ON VENUS
CAME TO BE
"PROOF" OF THE
SUN-CENTERED
"COPERNICAN"
VIEW OF THE
HEAVENS

AND, AS TIME WENT ON,
THE WORLD BEGAN TO
ACCEPT THE IDEA OF
THE EARTH AND ALL
THE PLANETS
REVOLVING AROUND
THE SUN.

WHICH WAS AT REST
AT THE CENTER OF
THE UNIVERSE.

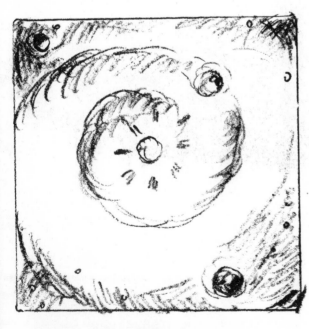
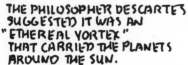
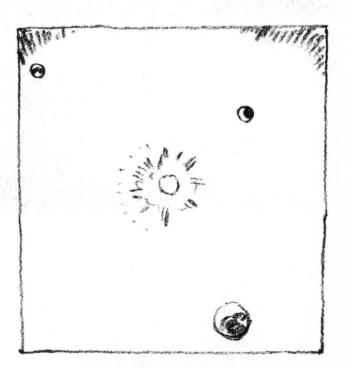

THE PHILOSOPHER DESCARTES
SUGGESTED IT WAS AN
"ETHEREAL VORTEX"
THAT CARRIED THE PLANETS
AROUND THE SUN.

IN THE LATE 17TH CENTURY,
ISAAC NEWTON PRESENTED
HIS THEORIES . .
AND THE FORCE OF GRAVITY
WAS ESTABLISHED AS THAT
WHICH HOLDS THE
UNIVERSE TOGETHER.

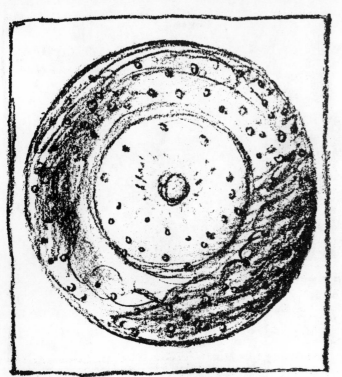

IN THE 18TH CENTURY THOMAS WRIGHT PROPOSED THE MILKY WAY UNIVERSE TO BE A SPHERICAL SHELL OF STARS.. AT THE CENTER OF WHICH WAS THE "EYE" OF THE SUPERNATURAL

IN A LATER MODEL, WRIGHT PICTURED THE MILKY WAY AS A SET OF CONCENTRIC RINGS OF STARS LIKE THE RINGS OF SATURN.. WITH AN EMPTY CENTER FOR THE SUPERNATURAL

THROUGHOUT THE 1780's, WILLIAM HERSCHEL OBSERVED THE MOVEMENTS OF TENS OF THOUSANDS OF STARS..

(IN THE LATER YEARS, WITH A 40 FOOT TELESCOPE HE ALSO BUILT)

CONCLUDING THAT THE SUN WAS NOT FIXED AT THE CENTER OF IT ALL.. BUT.. THE SUN WAS MOVING THROUGH SPACE!

AND THE "MILKY WAY" SEEMED TO BE SHAPED LIKE A GREAT POWDER PUFF

WITH THE SUN MOVING SOMEWHERE NEAR THE CENTER OF IT

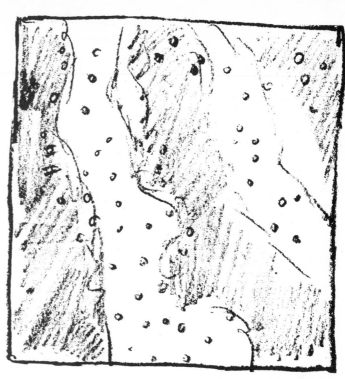

A ROTATING CIRCULAR ARRAY OF STARS WAS PICTURED BY IMMANUEL KANT A LITTLE LATER

A SIMILAR "PICTURE" CAME TO THE ASTRONOMER J.D. LAMBERT AT JUST ABOUT THE SAME TIME

THE MUSICIAN/ASTRONOMER WILLIAM HERSCHEL, IN 1781, DISCOVERED A NEW PLANET.. URANUS..
(WITH A 20 FOOT TELESCOPE HE BUILT HIMSELF)

AND DISCOVERED THAT THE MILKY WAY WAS MADE OF MILLIONS OF STARS!

OHN HERSCHEL (THE SON F WILLIAM) SPENT YEARS BSERVING THE SOUTHERN KIES.. AND THE AGELLANIC CLOUDS

IN 1833, HE FIRST SUGGESTED THERE COULD BE ANOTHER SEPARATE SYSTEM OF STARS SIMILAR TO THE MILKY WAY..
THE SYSTEM "M51"

WILLIAM PARSONS, WITH HIS NEW 72" REFLECTING TELESCOPE.. CONFIRMED HERSCHEL'S FINDINGS.. AS M51 COULD NOW BE SEEN AS A GREAT SPIRAL SYSTEM OF STARS.. WITH A BRIGHT CENTER

BEGINNING WITH THIS GREAT TELESCOPE, ASTRONOMERS COULD NOW SEE INDIVIDUAL STARS IN FAR AWAY NEBULAE

IN 1864, JOHN HERSCHEL COMPLETED THE "GENERAL CATALOGUE OF NEBULAE" STARTED BY HIS FATHER..
FATHER AND SON HAVING DISCOVERED 4,630 OUT OF THE 5,079 "OBJECTS" IN THE CATALOGUE

IN THE 17TH CENTURY
ISAAC NEWTON
HAD DISCOVERED
THAT SUNLIGHT
PASSING THROUGH
A "PRISM"..
SEPARATED INTO
LIGHT OF
ALL THE COLORS
OF THE RAINBOW

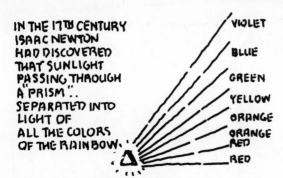

VIOLET
BLUE
GREEN
YELLOW
ORANGE
ORANGE
RED
RED

NOW, IN THE 19TH CENTURY,
ASTRONOMERS BEGAN
TO UNDERSTAND THE
"HAZY WISPS" OF
LIGHT IN THE SKIES
BY LOOKING THROUGH
THE PRISM OF A
"SPECTROSCOPE"

IN 1859,
THE PHYSICIST GUSTAV KIRCHOFF
AND THE CHEMIST ROBERT BUNSEN
IDENTIFIED ABOUT 24 KINDS
OF ATOMS IN THE ATMOSPHERE
OF THE SUN..
EACH TYPE OF ATOM PRODUCING
ITS OWN, UNIQUE "SPECTRUM"

IN THE 1860's, SIR WILLIAM HUGGINS
FOUND THAT THE PLANETARY NEBULAE
HAD LIGHT SPECTRUMS SIMILAR
TO GASES..   WHILE THE
"SPIRAL" NEBULAE HAD SPECTRUMS
SIMILAR TO THOSE OF THE ATOMS
OF STARS ✳'

IN 1912, THE ASTRONOMER
HENRIETTA LEAVITT, FROM
CAREFUL OBSERVATIONS
OF THE PULSATING LIGHT
OF THE "CEPHEID VARIABLE"
STARS OF THE SMALL
MAGELLANIC CLOUD...
DETERMINED A METHOD
TO MEASURE GREAT
DISTANCES IN SPACE !

AND THEN
THE ASTRONOMER
HARLOW SHAPLEY..
FROM OBSERVATIONS
OF THE GREAT NUMBERS OF
STARS NEAR SAGITTARIUS..
AND OBSERVATIONS OF
THE PULSATING LIGHT FROM
THE VARIABLE STARS WITHIN
THE GREAT GLOBULAR
CLUSTERS...

THE SUN

THE CENT OF THE GALA

CONCLUDED THAT
THE SUN
WAS NOT AT
THE CENTER
OF THE GALAXY..

BUT
NEARER
AN EDGE !

✳' THIS WAS THE BEGINNING OF THE UNDERSTANDING
THAT THERE WERE GREAT STAR SYSTEMS ..OR
"GALAXIES" OUTSIDE OUR OWN .. YEARS LATER..
IN 1912, THE ASTRONOMER VESTO SILPHER
BEGAN STUDYING LIGHT FROM WHAT WERE
CALLED AT THE TIME "SPIRAL NEBULAE"

VESTO SILPHER FOUND THAT 11 OUT OF 15
SPIRALS HE OBSERVED SHOWED
"REDSHIFTED" LIGHT FAR GREATER
THAN MOST OF THE STARS IN THE SKY...
OVER TIME.. HE FOUND MORE AND MORE
REDSHIFTED SPIRALS .. THE FAINTEST
APPEARING ONES SHOWING THE GREATEST
REDSHIFTS

IF THE REDSHIFTED LIGHT
FROM THE SPIRALS WAS
INTERPRETED AS A
DOPPLER TYPE REDSHIFT..
SOME SPIRALS WOULD BE
FLYING AWAY FROM US AT
685 MILES A SECOND !

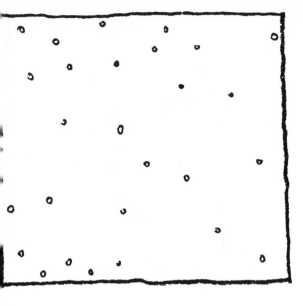

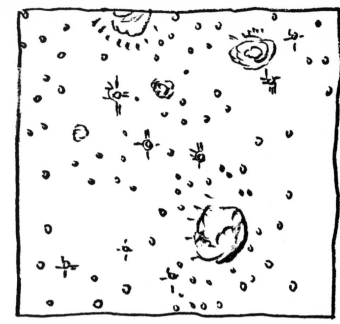

SINCE THE EARLY YEARS OF PHOTOGRAPHY, THERE HAD BEEN MANY ATTEMPTS TO PHOTOGRAPH THE MOON AND PLANETS AND STARS AND NEBULAE ..

DAGUERROTYPES AND "WET PLATE" PHOTOGRAPHS WERE USED WITHOUT TOO MUCH SUCCESS

IN 1879, THE ASTRONOMER HENRY DRAPER BEGAN TO PHOTOGRAPH THE HEAVENS THROUGH HIS TELESCOPES WITH A NEW, WORKABLE "DRY PLATE" PROCESS ...

ALLOWING ASTRONOMERS TO BEGIN TO SEE STARS WHERE NONE HAD BEEN SEEN BEFORE !

AROUND 1885, AT THE PARIS OBSERVATORY, THE HENRY BROTHERS BEGAN PHOTOGRAPHING REGIONS OF THE HEAVENS ..

WHICH INSPIRED THE BEGINNING OF A GREAT WORLD WIDE PROJECT .. TO PHOTOGRAPHICALLY "MAP" THE ENTIRE STAR SPHERE

AND THAT THE CENTER OF THE GALAXY IS IN A REGION OF THE SKY NEAR THE CONSTELLATION SAGITTARIUS

THE ASTRONOMER EDWIN HUBBLE AND MILTON HUMASON, BY 1936, HAD STUDIED MORE THAN 100 GALAXIES AND 10 CLUSTERS OF GALAXIES.

THE OBSERVED REDSHIFTS APPEARED TO INCREASE IN DIRECT PROPORTION TO THE GALAXIES DISTANCES

IF THE "DOPPLER" INTERPRETATION OF THE REDSHIFTS WAS CORRECT .. AND IF THE ASSUMED DISTANCES OF THE GALAXIES WAS CORRECT ..

THE MOST DISTANT GALAXIES THEY COULD SEE AT THE TIME WERE FLYING AWAY FROM US AT A SPEED OF 26,000 MILES A SECOND !

THE DISTANCE TO THE CENTER OF THE GALAXY IS THOUGHT TO BE ABOUT 32,000 LIGHT YEARS AWAY TODAY

UNTIL A LITTLE MORE
THAN 50 YEARS AGO,
IT WAS GENERALLY
THOUGHT THAT THE
WHOLE "UNIVERSE"
WAS OUR OWN
STAR SYSTEM

AROUND 1924, THE ASTRONOMER
EDWIN HUBBLE PROVED THAT
"THE ANDROMEDA SPIRAL"
AND OTHER SPIRALS
WERE ACTUALLY
BEYOND THIS ONE

THE DIM LIGHTS
OF THE UNIVERSE
HAD COME INTO VIEW..

SOME
WERE
CLOUD LIKE
NEBULAE

SOME WERE SYSTEMS OF
GREAT NUMBERS OF STARS
CALLED "ISLAND UNIVERSES"
LATER TO BE KNOWN AS
"GALAXIES"

THEIR NUMBERS
GREW OVER THE YEARS
AS THE TELESCOPES
IMPROVED
AND TODAY..
THERE ARE ESTIMATED
TO BE
BILLIONS OF GALAXIES!

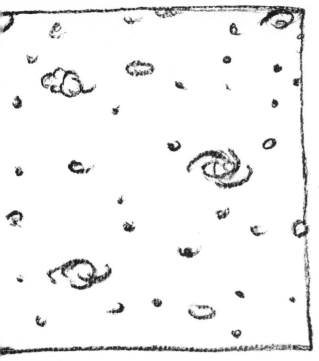

THAT THERE WERE
GREAT COMMUNITIES
OF STARS
FAR BEYOND
THIS GALAXY..

THE
MILKY WAY
"UNIVERSE"

CAME TO BE

THE
MILKY WAY
"GALAXY"

AROUND 1950
THE DETAILED SHAPE
OF OUR GALAXY
CAME TO LIGHT...

AND NOW
WE THINK OF IT
AS A GREAT
SPIRAL GALAXY

ABOUT
100,000 LIGHT
YEARS ACROSS

A COSMIC "WHEEL"
OF ABOUT
200 BILLION STARS

THE UNIVERSE

THROUGHOUT
MOST OF HISTORY
THE IDEA OF
" THE
UNIVERSE "
WAS BASED ON
ALL THAT COULD
BE SEEN AT THE
TIME · ·
ABOUT
A THOUSAND
STARS

THE IDEA OF
THE UNIVERSE
AS A SPHERE
HAS APPEARED
AT VARIOUS
TIMES
THROUGHOUT
HISTORY

SOME EARLY
PHILOSOPHERS
THOUGHT OF
THE UNIVERSE
AS A
SPHERE

AND THE
EARLY
ASTRONOMERS
IMAGINED
UNIVERSES
OF GREAT
CRYSTAL
SPHERES

WHAT IS THE UNIVERSE
OF TODAY ?

FROM THE TINIEST OF
THE TINY TO THE GRANDEST
OF THE GRAND · ·
WE LIVE IN ABOUT THE
MIDDLE OF IT ALL · · ·

THE TINIEST LEVEL
IS THAT OF THE
SINGULARITIES OF
THE OCEAN OF LIGHT · ·
THE BEGINNINGS OF
THE WAVES

THE LEVEL OF THE
QUARKS · ·
THE PHOTONS · ·
THE ELECTRONS ·

THE LEVEL OF THE
NUCLEUS · ·
THE PROTON
AND THE NEUTRON

THE LEVEL OF THE
ATOM

THE LEVEL OF THE
MOLECULE

THE LEVEL OF THE
LIVING CELL

THE LEVEL OF
LIFE
ON
EARTH

THE LEVEL OF
THE STARS

THE GALAXIES

THE GALAXIES
OF
GALAXIES

THE GALAXIES
OF
GALAXIES
OF
GALAXIES

THE
COSMIC

BANG

ONE GALAXY OF BILLIONS OF GALAXIES

THE MOST
POPULAR
"UNIVERSE"
THEORY OF
THE 1970's
WAS THE
"BIG BANG"
THEORY

IT TOLD OF
A UNIVERSE
THAT CAME
TO BE
15 TO 20
BILLIONS
OF YEARS
AGO

SOMETHING
ABOUT THE
SIZE OF
A SCOOP OF
ICE CREAM · ·
· · OR · ·
SOME SAY ·
· · THE SIZE OF
AN ATOM · ·
WENT
· BANG ·

HOW IT
GOT THERE
IN THE
FIRST
PLACE
AND
WHAT
MADE IT
GO
BANG · ?
NO ONE
KNOWS

THERE WAS
A TREMENDOUS
EXPLOSION
WHICH CAUSED
ALL OF SPACE
TO FLY
OUTWARD · ·

AND THE
UNIVERSE · ·
OF BILLIONS
OF GALAXIES · ·
IS SAID TO BE
STILL
ZOOMING
OUTWARD · ·

AT AN
INCREDIBLE
SPEED · ·
LIKE A
HUGE
GLOWING
BALLOON !

✳ THE THEORIES OF RELATIVITY
ARE THE FOUNDATION FOR MOST
OF THE IDEAS ABOUT THE
ORIGIN AND STRUCTURE
OF THE UNIVERSE OF TODAY

IN 1905,
THE PHYSICIST ALBERT EINSTEIN
PRESENTED THE
"THEORY OF RELATIVITY"

IT STATED THAT
TIME AND SPACE
ARE NOT SEPARATE IDEAS

THAT OUR PERCEPTION
OF TIME AND SPACE
IS DIFFERENT ON EARTH
THAN MOVING AT
GREAT SPEEDS
THROUGHOUT SPACE · ·

THAT ALL MOTION IS RELATIVE TO THE OBSERVER
· · MOTION, TIME AND DISTANCE ARE RELATIVE TO
MOVING FRAMES OF REFERENCE EXCEPT FOR
THE VELOCITY OF LIGHT IN SPACE · ·
WHICH IS A "CONSTANT" · ALWAYS THE SAME · · ·
INDEPENDENT OF THE VELOCITY OF THE OBSERVER
OR THE VELOCITY OF THE LIGHT SOURCE · · ·

AND · · · ·
MASS CAN BE CH
INTO ENERGY · · ·
AND ENERGY CA
BE CHANGED
INTO MASS

$E = mc^2$

THE EARLY YEARS OF THE 20TH CENTURY MANY NEW UNIVERSE IDEAS APPEARED

FOUNDED UPON THE THEORIES OF RELATIVITY* AND

THE OBSERVATIONS OF THE LIGHT OF THE STARS

THE MIDDLE OF THE 19TH CENTURY. THE MATHEMATICIAN GEORGE REIMANN PRESENTED A 4-DIMENSIONAL "HYPERSPHERE" UNIVERSE

IN 1917. THE ASTRONOMER WILHELM DE SITTER PRESENTED AN EVER EXPANDING HYPERSPHERE .. EXPANDING INFINITELY TOWARD "FLAT SPACE"

WHEREIN A RAY OF LIGHT WOULD MOVE IN AN ETERNALLY EXPANDING SPIRAL

IN THE 1920'S, THE SCIENTIST ALBERT FRIEDMANN PRESENTED SOLUTIONS WHEREIN THE UNIVERSE MIGHT EITHER EXPAND FOREVER .. OR . SOMEDAY COME BACK INWARD TO COLLAPSE

THE THEORISTS ARTHUR EDDINGTON AND GEORGE LEMAITRE PRESENTED THE IDEA OF AN EXPANDING UNIVERSE BEGINNING FROM A TINY PRIMORDIAL SPHERE

ABOUT 1927. "L'ATOME PRIMITIVE" THEORY WAS PRESENTED BY ABBE LEMAITRE .. TELLING OF AN "ATOM" OF ORIGIN FLASHING FORTH 50 BILLION YEARS AGO ..

TEN BILLION YEARS AGO .. ARRIVING AT A SPHERE .. 2 MILLION LIGHT YEARS ACROSS TO BEGIN TO FORM THE GALAXIES OUT OF THE CLOUDS

WHO REALLY KNOWS HOW THIS UNIVERSE CAME TO BE?

WHO WAS THERE AT THE TIME?

BRIGHT BLUE SKIES

IN THE LAST FEW YEARS .. ALL OVER THE WORLD

A GREAT ARRAY OF NEW IDEAS HAVE COME TO LIGHT ..

AS HAVE NEW VERSIONS OF SOME OF THE OLDER IDEAS

IN A SPRINGTIME OF COSMIC DREAMS

1916, ALBERT EINSTEIN PRESENTED "GENERAL THEORY OF RELATIVITY".. BROADENING THE ORIGINAL THEORY TO INCLUDE GRAVITATION AND ACCELERATION

STATING THAT OUR PERCEPTION OF TIME AND SPACE IS DIFFERENT IN THE STRONG GRAVITATIONAL FIELD OF EARTH THAN IN THE LIGHTER GRAVITATIONAL FIELD OF SPACE ... AND ..
THE UNIVERSE WE PERCEIVE IS DEPENDENT UPON OUR RELATIVE ACCELERATIONS ..
DISTANCE AND TIME ARE NOT ABSOLUTE!

IN HIS "FIELD EQUATIONS" HE ASSUMED THAT THE UNIVERSE, OVERALL. CAN BE REGARDED AS BEING THE SAME DENSITY THROUGHOUT.. AND THAT..

THE PROPERTIES OF THE UNIVERSE CAN BE REGARDED TO BE GENERALLY "THE SAME" THROUGHOUT THE UNIVERSE.

HERE IS THE STORY
OF THE BEGINNING
OF THE STORY
OF
"THE BIG BANG"

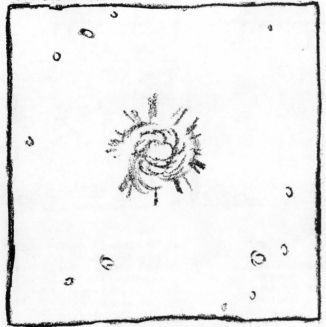

STARS MOVING
AWAY FROM US
WOULD APPEAR
REDDER
AS THE LIGHT WAVES
IN A DOPPLER EFFECT
APPEAR TO LENGTHEN
TOWARD THE RED
END OF THE VISIBLE
LIGHT SPECTRUM

IN THE 1920's
THE ASTRONOMER
EDWIN HUBBLE
SHOWED THERE
WERE A GREAT
MANY GALAXIES
OUTSIDE OURS..

AND THAT THE LIGHT
COMING FROM
THEM WAS
"REDSHIFTED"

THIS WAS
TAKEN TO MEAN
THAT THE LIGHT
WAVES WERE
APPEARING
LONGER..

..MORE
"STRETCHED OUT"
THE FASTER THE
STAR OR GALAXY
WAS MOVING
AWAY FROM
US

THIS LED SOME
TO CONCLUDE
THAT THE
UNIVERSE
IS
EXPANDING

ACCORDING TO THE
EXPANDING UNIVERSE
THEORY..
THE GALAXIES
AND THE EARTH
ARE FLYING
APART AT
GREAT SPEEDS...

EVERY POINT
IN THE UNIVERSE
IS MOVING AWAY
FROM EVERY
OTHER POINT
WITH A SPEED
PROPORTIONAL
TO THE DISTANCE
BETWEEN THE
POINTS

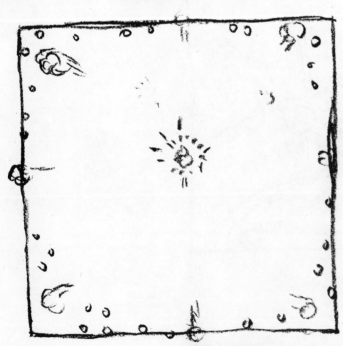

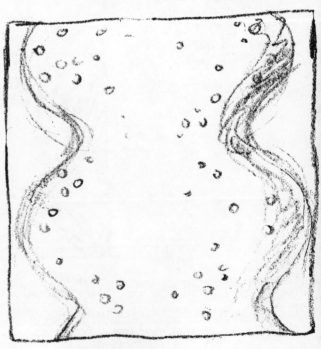

SOME SAID THAT
THE UNIVERSE IS
ETERNALLY EXPANDING
AND THERE IS
A CONTINUAL CREATION
OF NEW MATTER
TO TAKE THE PLACE
OF THE GALAXIES
THAT ARE ZOOMING
OUTWARD)...

SOME SAID THAT
THE UNIVERSE IS
EXPANDING NOW..
IN A PHASE THAT
OSCILLATES BETWEEN
EXPANSION AND
CONTRACTION..
AND THERE'S A
FIXED AMOUNT
OF MATTER

THE EARTH AND
THE GALAXY WE LIVE IN
ARE MOVING IN
SPACE AT ABOUT
A MILLION MILES
AN HOUR ! (?)

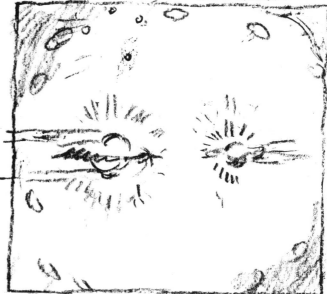

STARS MOVING TOWARD US
WOULD APPEAR "BLUER"
AS THE LIGHT WAVES..
IN A DOPPLER EFFECT..
APPEAR TO SHORTEN
TOWARD THE ULTRAVIOLET
END OF THE VISIBLE
LIGHT SPECTRUM

ACCORDING TO
THE EXPANDING
UNIVERSE THEORY

THE FURTHER
AWAY A GALAXY
IS FROM US,
THE FASTER
IT'S MOVING !

AND EVERY
GALAXY IN THE
UNIVERSE
IS FLEETING
AWAY FROM
EVERY OTHER
GALAXY

EXCEPT FOR
LOCAL GALAXIES
WITHIN GROUPS..
OR STARS WITHIN
GALAXIES
THAT COULD BE
MOVING TOWARD
EACH OTHER
GRAVITATIONALLY

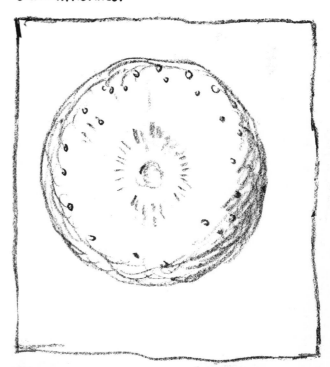

SOME SAID
THAT IF THE UNIVERSE
IS NOW EXPANDING,
IT MUST HAVE BEEN
CLOSER AND CLOSER
TOGETHER
GOING FURTHER
AND FURTHER
BACK INTO TIME...

IF THE UNIVERSE
WAS ON A MOVIE FILM
RUN IN REVERSE..
THE UNIVERSE
WOULD BECOME
LITTLER AND LITTLER..
THE GALAXIES WOULD
RUSH CLOSER
AND CLOSER
TOGETHER..

ALL THE
STARS AND GALAXIES
WOULD BECOME
ONE GLOWING
FIREBALL
BECOMING LITTLER
AND LITTLER..
HEATING UP
TO TRILLIONS
OF DEGREES

THE SIZE OF
A TINY BALL
THAT COULDN'T
GET ANY
TINIER...
NOW.. RUN THE MOVIE
IN "FORWARD"
AND YOU HAVE
THE STORY OF
"THE BIG BANG"

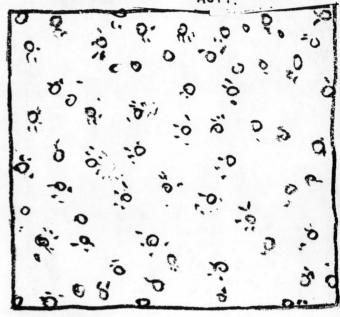

AFTER ONE BILLION BILLION BILLION BILLION BILLIONTH OF A SECOND..

AN INCREDIBLE FIREBALL.. 10,000,000,000, 000,000,000, 000,000,000 °C. HOT!!

BANG

AN EXPLOSION OF SPACE ITSELF.. HAPPENING SIMULTANEOUSLY EVERYWHERE.. EVERY PART RUSHING AWAY FROM EVERY OTHER PART!

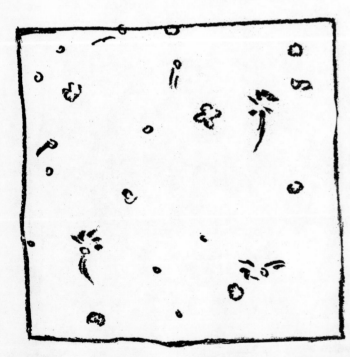

FOR THE FIRST 2,000 YEARS MORE LIGHT THAN MATTER

CONTINUING OUTWARD,... AFTER 100,000 YEARS. ATOMS BEGIN TO HOLD TOGETHER

BUT THEY ONLY LAST FOR HUNDREDS OR THOUSANDS OF YEARS.. BECAUSE THE "ORIGINAL" HIGH ENERGY PHOTONS KEEP FREEING THE ELECTRONS FROM ORBIT

UNTIL..THE ORIGINAL PHOTONS LOSE ENERGY OVER TIME.. AND BEGIN TO PASS AMONG THE ATOMS WITHOUT DISRUPTING THEM.

AFTER SEVERAL HUNDREDS OF THOUSANDS OF YEARS.. THE HYDROGEN AND THE HELIUM ATOMS LAST

FOR HUNDREDS OF MILLIONS OF YEARS.. THE "BALLOON" CONTINUES TO ZOOM OUTWARD

UNTIL SOME ATOM CLUSTER TOGETHER TO BEGIN

THE CLOUDS WITHIN THE CLOUDS

NO LIGHT MATTER AT A TEMPERATURE TEN MILLION
OF ONE TRILLION TIMES DENSER
DEGREES, THAN ATOMS..
THE UNIVERSE A SPHERE OF
IS ALL RADIATION.. PURE ENERGY

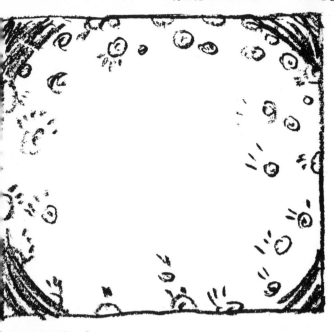

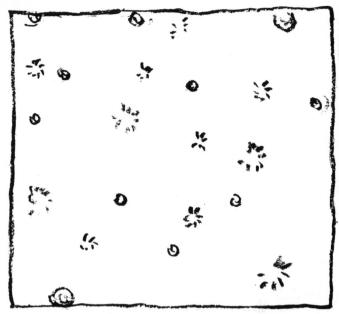

THE UNIVERSE
ZOOMS OUTWARD..

FLASHING FORTH
THE FIRST
WHITE HOLES!

COOLING   AFTER ABOUT   A UNIVERSE
INTO   30 MINUTES   OF MOSTLY
A   IT'S COOL ENOUGH   -LIGHT:...
UNIVERSE   FOR SOME   PHOTONS
OF   ATOMIC "NUCLEI"   NEUTRINOS
"QUARKS"   TO HOLD   PROTONS
   TOGETHER   NEUTRONS
   ELECTRONS

...LLION YEARS   CONTINUING   WHICH SOMEDAY   TO LIGHT   AS A STAR   AND NOW.   WITH
...ER THE   TO ZOOM   DISAPPEAR   NEW STARS..   -- A CLOUD   15 TO 20 BILLION   ABOUT
..."BANG"   OUTWARD..   TO RETURN   TO LIVE   -- A STAR   YEARS AFTER   A HUNDRED
...HE   AS   THEIR DUST   TIME   -- A CLOUD   THE "BANG"   BILLION
...LAXIES   THE STARS   TO THE   AND   -- A STAR   THE UNIVERSE   STARS
...BEGIN   BEGIN   CLOUDS..   TIME   -- !!!!!   HAS LIT ITSELF   IN EACH
   TO   AGAIN   --   WITH   GALAXY!!
   "LIGHT"   A HUNDRED
   "!!!   BILLION
         GALAXIES.

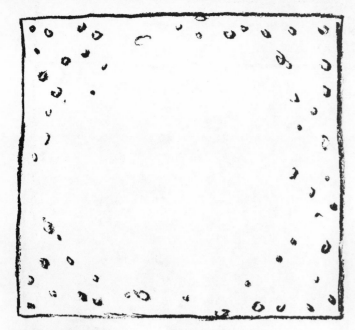

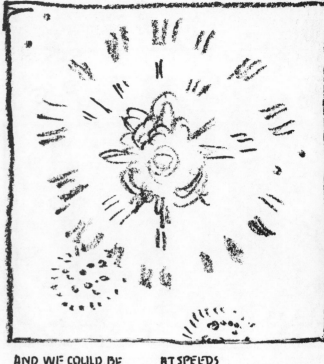

IN THE BIG BANG
EXPANDING
UNIVERSE
VIEW..

SPACETIME <u>ITSELF</u>
IS ZOOMING
OUTWARD
LIKE AN
EXPANDING
RUBBER BALLOON !

AND
THE GALAXIES
THEMSELVES
ARE MOVING
OUTWARD
LIKE EXPANDING
POLKA DOTS
ON AN
EXPANDING
BALLOON

AND WE COULD BE
ZOOMING AWAY
FROM SOME
QUASARS
THAT COULD BE
UP TO
15 BILLION
LIGHT YEARS
AWAY

AT SPEEDS
OF UP TO
90%
OF
THE SPEED
OF
LIGHT !

DOES THE UNIVERSE GO ON FOREVER

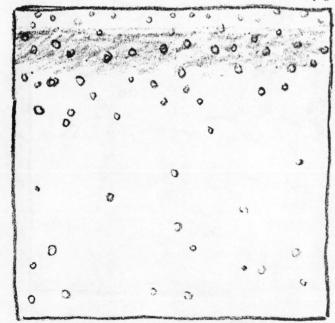

IN THE VIEW OF
MATHEMATICIANS
THERE ARE
OTHER KINDS OF
SPACE..

THE EXPANDING
UNIVERSE IS THOUGHT
TO BE

(A)
NEGATIVELY CURVED..
"OPEN, INFINITE"
SPACE..
LIKE ORDINARY
SPACE..
HAS NO LIMITS
AND GOES ON
FOREVER..
         ..OR..

(B)
POSITIVELY CURVED
"CLOSED" SPACE..
"FINITE"..WHEREIN
THE UNIVERSE
HAS LIMITS
AND LOOKS LIKE
A SPHERE

IS THE UNIVERSE
INFINITE..
   "OPEN"
      SPACE ?
DOES IT GO
ON FOREVER ?

ORDINARY, "EUCLIDIAN"
SPACE IS CONSIDERED
TO BE INFINITE.."OPEN"..
SPACE THAT GOES ON
FOREVER

   ..JUST AS IT
      APPEARS
      TO MOST
      OF US

SOME SAY
THE UNIVERSE
IS "OPEN"..

AND THERE ISN'T
ENOUGH MATTER
IN THE UNIVERSE
TO EVER STOP
THE EXPANSION

AND IT'LL KEEP
GOING UNTIL
SOME DAY WHEN
IT BECOMES
NOTHING BUT
WIFFS AND
PUFFS..AND
THEN
NOTHING AT ALL !

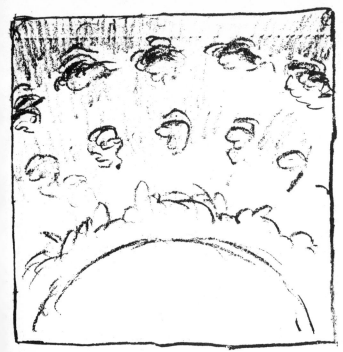

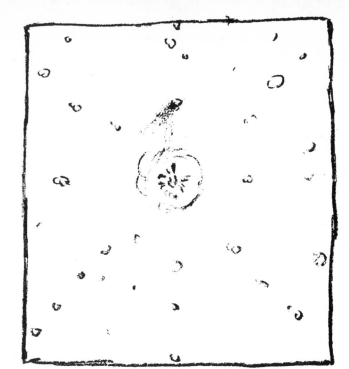

SOMEDAY SOON WE SHOULD BE ABLE TO SEE BILLIONS OF LIGHT YEARS FURTHER BACK INTO TIME

IF THE BIG BANG HAPPENED ACCORDING TO THE THEORY

WE SHOULD BE ABLE TO SEE THE BEGINNINGS OF THE GALAXIES

AND, PERHAPS, THE FIRST ATOMS EMERGING FROM THE LIGHT

IN A POSITIVELY CURVED, SPHERICAL, "CLOSED" SPACE UNIVERSE .. AN ASTRONOMER WITH AN IMMENSELY POWERFUL TELESCOPE WOULD EVENTUALLY BE GAZING AT THE BACK OF HIS OWN HEAD !

OME SAY HE UNIVERSE S "CLOSED".. ND THERE IS NOUGH MATTER N THE UNIVERSE ALTHOUGH WE MAY NOT HAVE EEN IT ALL YET)

AND THE UNIVERSE WILL COME TO REST SOMEDAY.. AND THEN BEGIN TO FALL BACK WITHIN ITSELF !

RECENTLY, GREAT HALOS HAVE COME INTO VIEW AROUND GALAXIES .. WHICH, SOME SAY, COULD BE ENOUGH MATTER TO "CLOSE" THE UNIVERSE

IF THE UNIVERSE IS "CLOSED", ACCORDING TO SOME, AFTER ABOUT 40 BILLION YEARS THE UNIVERSE WOULD SLOWLY COME TO REST AND THEN BEGIN TO FALL BACK WITHIN ITSELF

FALLING FASTER AND FASTER WITHIN .. UNTIL .. 80 BILLION YEARS AFTER THE BIG BANG BANG,. ANOTHER . BANG. . . . .

WHAT GOES ON IN THE SINGULARITY ?

NOTHINGNESS ?

IS A NEW UNIVERSE BORN ?

SOME SAY THAT EVEN IF THE ENTIRE UNIVERSE TURNED INTO A SINGULARITY

IT COULD STILL EXIST IN THE QUANTUM REALM

A TRUE SINGULARITY NEED NOT HAVE EXISTED AT THE BEGINNING OF THIS UNIVERSE ACCORDING TO THE MATHEMATICIA, G.F.R. ELLIS

SOME SAY THAT EVEN IF THE UNIVERSE IS EXPANDING ..

THERE NEED NOT HAVE BEEN A "BIG BANG"

IN WHAT IS KNOWN AS THE "WHIMPER COSMOLOGY" .. A ONE AND ONLY SINGULARITY NEED NOT EXIST AT A UNIVERSE BEGINNING

EVEN THOUGH INFINITE PRESSURE AND TEMPERATURE EXIST AT THE TIME OF THE SINGULARITY

AS "WELL BEHAVED" MATTER CAN ENTER THE UNIVERSE . THROUGH A "CAUCHY HORIZON"

A "CAUCHY HORIZON" APPEARS TO ACT SOMETHING LIKE A SINGULARITY ..

BUT IT IS NOT A TRUE SINGULARITY

SOME SCIENTISTS OF TODAY THINK THERE ARE WAYS FOR ENERGY TO ENTER THIS UNIVERSE FROM ANOTHER UNIVERSE

ONE THEORY TELLS OF A BORDERING "NEGATIVE" UNIVERSE .. OR ANTIGRAVITY UNIVERSE .. WITH AN INFINITE NUMBER OF ADJACENT POSITIVE AND NEGATIVE UNIVERSES ..

WHEREIN THE OBSERVED "COSMIC BACKGROUND RADIATION" IS STARLIGHT SHINING THROUGH THE "TIMELESS BOUNDARY" .. FROM AN ADJACENT UNIVERSE.

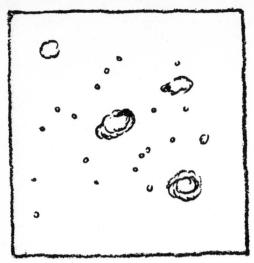

SOME SAY
THAT EVEN IF
THE WHOLE
UNIVERSE
TURNED INTO
A "BLACKHOLE"

IT COULD STILL
EXIST
BECAUSE
IT WOULD BE
SO BIG
THAT THE DENSITY
WOULD BE
RELATIVELY
LIGHT

SOME SAY
THE UNIVERSE
OSCILLATES
BETWEEN
"BIG BANGS"
THAT GO
ON FOREVER

BANG..
EXPAND..
CONTRACT..
BANG:
ETC. ETC.

AND
EACH TIME
OUT OF THE
SINGULARITY
THE NEXT
UNIVERSE
IS DIFFERENT

CAUSE OF THE
ME LIKE NATURE
F THE SINGULARITY

A STAR THAT
IS BECOMING
A BLACKHOLE
IN ONE PLACE
COULD BE
EMERGING
AS A WHITEHOLE
IN ANOTHER

PERHAPS
MATTER ENTERS
OUR UNIVERSE
THROUGH
SPACE-TIME
"BRIDGES"
OUT OF
THE
SINGULARITY

APPEARING
AS
LIGHT..
FROM
WHITEHOLES?

OR
QUASARS
?

IT COULD
ORIGINATE
IN ANOTHER
COEXISTANT
UNIVERSE

.. IT COULD
ORIGINATE
FROM ANOTHER
TIME AND SPACE
WITHIN
THIS UNIVERSE!

THE "STEADY STATE" THEORIES HAVE NOT BEEN AS POPULAR AS THE BIG BANG THEORY IN RECENT YEARS..

BUT THE IDEA OF AN ETERNAL UNIVERSE IS REESTABLISHING ITSELF IN THE LIGHT OF SOME AMAZING NEW DISCOVERIES AND SPECULATIONS

THE "STEADY STATE" THEORY TELLS OF AN EXPANDING, YET ETERNAL UNIVERSE

A UNIVERSE WHEREIN MATTER IS CONTINUALLY CREATED AS THE UNIVERSE EXPANDS OUTWARD

IN THE ORIGINAL VERSION OF THE STEADY STATE, NEW MATTER CONTINUALLY APPEARS

UNIVERSALLY THROUGHOUT.. AS ATOMS OF HYDROGEN

IN A LATER VERSION OF THE STEADY STATE THEORY

MATTER IS STILL CONTINUOUSLY CREATED

THE STEADY STATE THEORY WAS PRESENTED BY FRED HOYLE.. THROUGH THE YEARS HE PRESENTED A NUMBER OF VARIATIONS OF IT.

FOR YEARS, THE "BIG BANG" AND THE "STEADY STATE" WERE THE TWO MOST POPULAR COSMOLOGICAL THEORIES

THE "BIG BANG" THEORY WAS DEVELOPED AND CHAMPIONED OVER THE YEARS BY GEORGE GAMOW

THE NAME "BIG BANG" WAS HUMOROUSLY GIVEN TO THE THEORY BY FRED HOYLE

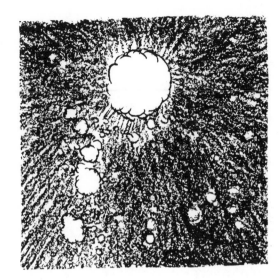

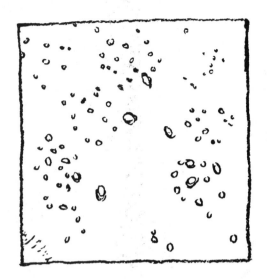

AS "NEW" ATOMS AND
ANCIENT ATOMS
COME TOGETHER
TO BEGIN TO FORM
CLOUDS
WHEREIN THE STARS
ARE BORN

TO SOMEDAY COME TO LIGHT
AS NEW GALAXIES
OF BRIGHT YOUNG STARS
TO TAKE THE PLACE OF
THE OLDER GALAXIES
THAT HAVE ZOOMED
OUTWARD

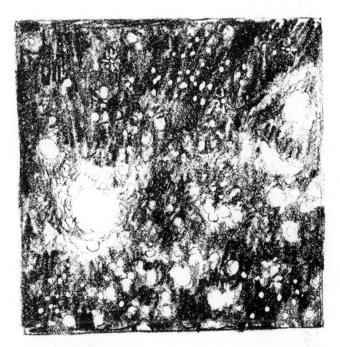

WHERE
NEW LIGHT
SPRINGS FORTH
FROM THE DUST
OF THE OLD

NOW AT CONCENTRATED
HIGH DENSITY REGIONS
LIKE QUASARS...OR
BLACK HOLES...OR
WHITEHOLES

AN ETERNAL BALANCE
OF OLD AND NEW..
A CONTINUAL CREATION
OF LIGHT

THE BIG BANG IDEA
IS ONLY ONE OF
A VAST NUMBER
OF CONCLUSIONS
THAT COULD BE
MADE FROM THE
SAME OBSERVATIONS !

EVEN IF THE
UNIVERSE IS
EXPANDING
(IT MAY NOT BE !)

IT NEED
NOT HAVE
HAD A
BEGINNING
AT ALL !

EVEN IF IT
HAD A
BEGINNING,
IT NEED NOT
HAVE STARTED
FROM A
TINY SPHERE
!

ONE RECENT THEORY
SUGGESTS THAT THE
UNIVERSE IS JUST AS
IT APPEARS ·· ·
TOTALLY AT REST...

AND THE MIND-BOGGLING
IDEAS OF THE THEORY
OF RELATIVITY
RELATE ONLY
TO LOCAL CONDITIONS,
WHERE THE GRAVITATIONAL
FORCES ARE QUITE STRONG

JUST AS THERE ARE
DIFFERENT GOVERNING
FORCES FOR ATOMS ··
AND FOR ELEMENTARY
PARTICLES ·· AND FOR
THE STARS AND GALAXIES...

THERE MAY BE
DIFFERENT GOVERNING
FORCES FOR THE
STARS AND GALAXIES
(GRAVITY) ·· AND,
FOR SOMETHING OF
A GREAT COSMIC ORDER ··
THE UNIVERSE ITSELF !

A "THREE-UNIVERSE"
THEORY CALLS FOR A
FORWARD-IN-TIME
REGULAR UNIVERSE
COEXISTING WITH A
BACKWARD-IN-TIME
ANTIMATTER UNIVERSE ··

TOGETHER WITH A
"MIRROR" UNIVERSE
OF FASTER THAN LIGHT
"LIGHT"··

(J·R·GOTT)

ANOTHER THEORY
TELLS OF A CLOSED
SPACE-TIME WHEREIN
THE UNIVERSE IS
OSCILLATING IN
CYCLES ··
FORWARD IN TIME ··
AND
BACKWARD IN TIME !

(P. DAVIES)

THE UNIVERSE GETS
OLDER IN A FORWARD-
IN-TIME "MATTER"
UNIVERSE ·· AND
REJUVINATES IN A
"BACKWARD-IN-TIME"
ANTI-MATTER UNIVERSE ··
IN A SYMETRICAL
"DOUBLE-LOOP" CLOSED
SPACE-TIME !

THE SUN AT THE CENTER ↓

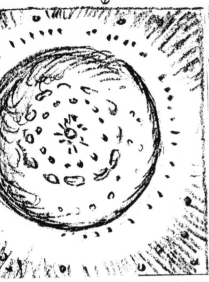

A LOT OF BUBBLES

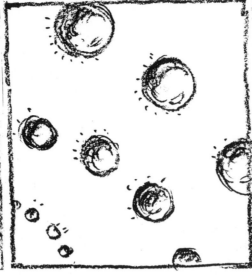

ONE THEORY SUGGESTS THAT OUR SUN IS AT THE CENTER OF A "METAGALAXY"

AN IMMENSE SUPER SYSTEM OF GALAXIES THAT IS BILLIONS OF LIGHT YEARS ACROSS..

THAT IS NOT THE WHOLE UNIVERSE

AN ENDLESSLY PULSATING BALANCE OF MATTER AND ANTIMATTER..

ALWAYS AT LEAST BILLIONS OF LIGHT YEARS ACROSS..

THAT CAME FROM AN ORIGINAL SPHERE A TRILLION LIGHT YEARS ACROSS

(ALFVEN AND KLEIN)

ANOTHER TELLS OF A GREAT NUMBER OF COEXISTANT, EXPANDING, OPEN "BUBBLE" UNIVERSES

(J.R. GOTT)

ANOTHER TELLS OF POSSIBLE "OSCILLATING ISLANDS" IN THE OCEAN OF SPACE..

LOCAL REGIONS OF SPACE EXPANDING OR CONTRACTING IN VARIOUS SIZES AND AMPLITUDES..

WITHIN A UNIVERSE "AT REST"

(HOYLE)

THE EARTH AT THE ↙ CENTER

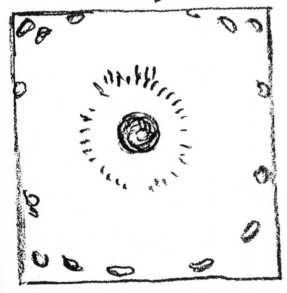

THE COSMIC REJUVINATOR

ONE RECENT THEORY PRESENTS THE EARTH AT THE CENTER OF THE UNIVERSE..

AS IT WAS BEFORE THE TIME OF COPERNICUS!

..A UNIVERSE AT REST.. SPHERICAL AND SYMETRICAL ..WHEREIN SPACE ITSELF CURVES AROUND TO A SINGULARITY AT THE "ANTICENTER"

..AN EVERPRESENT "COSMIC REJUVINATOR" WHERE THE ANCIENT GALAXIES GO TO.. AND WHERE NEW HYDROGEN IS SENT FORTH..

(G.F.R. ELLIS)

..AN EVERPRESENT "SOURCE" WHEREIN CREATION IS CONTINUAL!

THE UNIVERSE AT REST

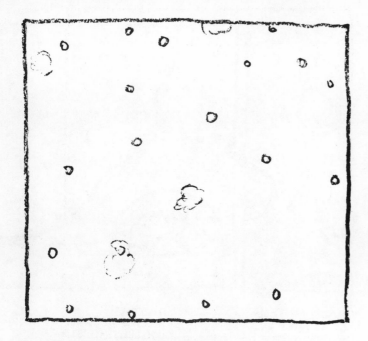

SOME THEORIES CALL FOR A
UNIVERSE THAT IS TOTALLY STILL...

BY REINTERPRETING THE IDEA
OF WHAT THE REDSHIFTS MEAN..
THE UNIVERSE IS NOT EXPANDING AT ALL!

PERHAPS THE UNIVERSE IS AT REST..
ETERNAL..

JUST AS IT LOOKS!

THERE ARE THE
"CHANGING MASS"
COSMOLOGIES

WHEREIN THE
MASSES OF
FUNDAMENTAL
PARTICLES
CHANGE
OVER TIME..
CAUSING A
CHANGE
IN THE
WAVELENGTH
OF LIGHT
THEY EMIT

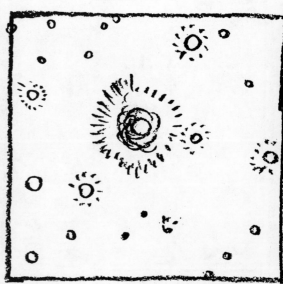

PERHAPS THE LIGHT
FROM OTHER GALAXIES
APPEARS
IN LONGER WAVELENGTHS
BECAUSE
"NEWLY CREATED" MATTER
PRODUCES
LONGER WAVELENGTHS

NEWLY CREATED MATTER
COULD BE OF LESSER
MASS.. WITH WIDER
ELECTRON ORBITS
ABSORBING AND
EMITTING LONGER
WAVELENGTHS THAN
OLDER MATTER

NEWLY CREATED MATTER
OVER TIME, WOULD BE
INFLUENCED BY THE
MORE ANCIENT ATOMS
ALL OVER THE UNIVERSE
..AND GRADUALLY
INCREASE IN MASS
TO BECOME THE SAME
AS THE OTHERS

WHAT APPEARS TO BE
GALAXIES SPEEDING
AWAY FROM EACH OTHER
COULD BE GALAXIES
"AT REST"..

THE "LIGHT YEAR"
RULER COULD BE
GETTING LITTLER
OVER TIME!

TIRED LIGHT

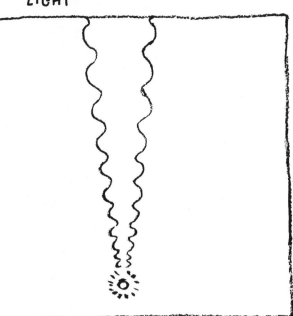

VARYING "G"

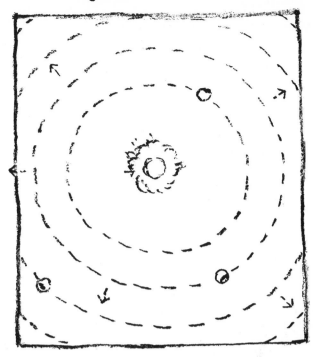

HERE ARE THE "TIRED LIGHT" COSMOLOGIES .. THAT INTERPRET THE REDSHIFTS AS LIGHT LOSING ENERGY OVER THE YEARS .. RATHER THAN A FLYING APART OF THE GALAXIES

THE EARLY "TIRED LIGHT" THEORIES SUGGESTED SOME UNKNOWN INTERACTION BETWEEN SOMETHING IN SPACE .. OR SPACE ITSELF ..

CAUSES LIGHT TO LOSE ENERGY AS IT TRAVELS GREAT DISTANCES

THERE ARE THE "VARYING G" COSMOLOGIES .. WHEREIN THE FUNDAMENTAL GRAVITATIONAL CONSTANT CHANGES WITH TIME

ONE RECENT "VARYING G" THEORY PRESENTED A UNIVERSE AT REST .. WITHIN WHICH THE PLANETS OF THE SOLAR SYSTEM ARE BALLOONING OUTWARD JUST A LITTLE AWAY FROM THE SUN.

(·DIRAC)

NEW MATTER CREATED OUT OF NOTHINGNESS ?

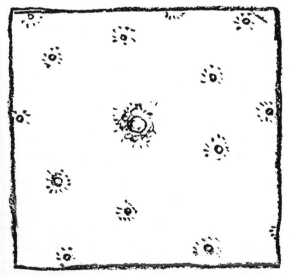

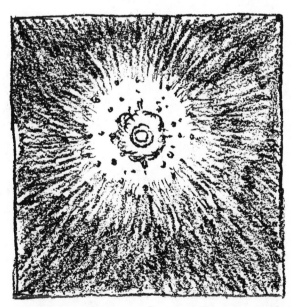

SOME OF THE LATEST "VARYING G" THEORIES ARE GENERALIZED TO INCLUDE THE IDEA OF CHANGING MASS .. AND OF MATTER APPEARING OUT OF NOTHINGNESS".

(DIRAC)

A RECENT THEORY OF "MULTIPLICATIVE CREATION" TELLS OF NEW MATTER BEING CREATED OUT OF "NOTHINGNESS" - WITHIN THE STARS THEMSELVES ?!

IF WE COULD "ZOOM·IN" TO LOOK UP CLOSE AT A LITTLE WHITEHOLE IN SPACE

WE WOULD SEE SOMETHING LIKE A CONTINUAL LITTLE BIG BANG

A CONTINUAL CREATION OF LIGHT

LASTING FOR THOUSANDS OF YEARS CREATING ·ELECTRON/POSITRONS ·PHOTONS, NEUTRINOS. ·····AND GRAVITONS !

A LOT OF
RECENT OBSERVATIONS
DON'T "FIT"
THE MOST POPULAR
"BIG BANG" STORY
OF TODAY

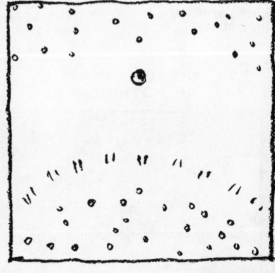

WITHIN A FEW YEARS
ASTRONOMERS WILL
KNOW MORE ABOUT
THE MYSTERIOUS
QUASARS . .

THERE ARE
THE UNEXPLAINED
GALAXY - QUASAR
ASSOCIATIONS

. . AND . . .

THERE IS THE
UNEXPLAINED
"FASTER - THAN - LIGHT"
GALAXY . .
AND THE
UNEXPLAINED
"FASTER THAN LIGHT"
QUASARS

IF QUASARS ARE NOT
AT THE EDGE OF
THE UNIVERSE . .
AND IF THEY'RE
NOT UNIFORMLY
DISTRIBUTED
THROUGHOUT
THE SKIES . . .

IT COULD MEAN
THAT THE
UNIVERSE
IS NOT
EXPANDING . . .

AND THERE
MAY NOT
HAVE BEEN
A "BANG"
BILLIONS OF
YEARS AGO !

ANOTHER
FOUNDATION
OF THE BIG BANG
EXPANDING
UNIVERSE
THEORY IS THE
"HUBBLE LAW"
DOPPLER TYPE
INTERPRETATION
OF THE REDSHIFTS . .

AND THERE'S A LOT
OF PROBLEMS
WITH THIS TOO . .

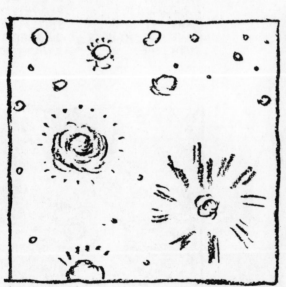

ON TOP OF
ALL OF THIS ,

IF THIS IS SO,

RECENT "REDSHIFT"
OBSERVATIONS
SHOW GALAXIES
VERY NEAR
EACH OTHER
WITH GREATLY
DIFFERING
"REDSHIFTS"

AND EVEN
REGIONS
WITHIN
SOME
GALAXIES
DIFFER
IN REDSHIFT
BY QUITE
A BIT !

PERHAPS MANY
OBSERVATIONS
ARE INACCURATE . .
OR PERHAPS . .
THE "DOPPLER"
INTERPRETATION
HAS LIMITED
VALIDITY

IN 1979, SOME
ASTRONOMERS SAID
THAT THE "BIG BANG"
UNIVERSE IS ONLY
9 BILLION LIGHT
YEARS ACROSS . .

NOT THE 15 TO 20
BILLION LIGHT YEARS
IT'S NORMALLY
THOUGHT TO BE . . .

STATING THAT
"HUBBLE'S CONSTANT"
(USED TO DETERMINE
DISTANCES IN SPACE)
DIDN'T TAKE INTO
ACCOUNT THE IDEA
THAT THE EARTH
ITSELF IS MOVING
TOWARD THE CENTER
OF IT'S OWN CLUSTER
OF STARS . .

ALL DISTANCES
SPACE ARE A LO
LESS THAN HAD
BEEN THOUGHT.
AND ANY BIG BAN
UNIVERSE WOUL
BE ABOUT
HALF AS BIG
AND ABOUT
HALF AS OLD
AS IT'S NOW
THOUGHT TO BE

THERE ARE OTHER WAYS
THAT LIGHT FROM THE
GALAXIES MIGHT APPEAR
LESS "BLUE" . . .

IF LIGHT WAVES
APPEAR
LONGER IN A STRONG
GRAVITATIONAL
FIELD

IF THE LIGHT IS APPROACHING
SO VERY SWIFTLY . . THE
NORMALLY INVISIBLE
INFRARED REGION
OF THE SPECTRUM IS
SHIFTED INTO THE VISIBLE

IF LIGHT LOSES A LITTLE ENERGY
(IN ENCOUNTERS WITH OTHER
PHOTONS ?) WHILE TRAVELING
IN SPACE

AND
. . . MORE

TODAY,
MOST OF THE
THEORIES OF
THE BEGINNING
OF THE
UNIVERSE
ARE FOUNDED
UPON
THE THEORY
OF
RELATIVITY

ISOUR GALAXY
ZOOMING TOWARD
THE SUPERGIANT
AT MORE THAN
A MILLION MILES
AN HOUR ?

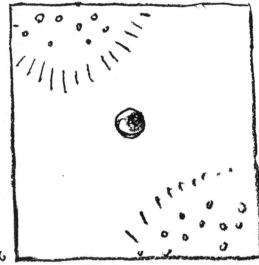

ONE OF THE FOUNDATIONS
OF THE BIG BANG EXPANDING
UNIVERSE THEORY IS THAT
THE UNIVERSE IS PRETTY MUCH
THE SAME "DENSITY"
THROUGHOUT · ·

YET · · MANY OBSERVATIONS
SHOW US A UNIVERSE
THAT MIGHT NOT BE
ESSENTIALLY ALIKE
THROUGHOUT

ACCORDING TO SOME
ASTRONOMERS,
"DISTANT" QUASARS
APPEAR TO BE LOCATED
MOSTLY NEAR THE
GALACTIC NORTH POLE · ·
AND IN ONE REGION
OF THE SOUTHERN
GALACTIC HEMISPHERE

THE MAGNITUDE AND VELOCITY
OF THE GALAXIES OF THE
NORTHERN HEMISPHERE SKIES
APPEAR TO BE DIFFERENT FROM
THOSE OF THE SOUTHERN SKIES · ·
(THESE OBSERVATIONS MAY BE
DUE TO OUR GALAXY SPEEDING
THROUGH SPACE TOWARD A
LOCATION IN PERSEUS)

RECENT (UNCONFIRMED)
OBSERVATIONS SHOW
THERE MIGHT BE A
SUPERGIANT CLUSTER
OF GALAXIES · ·
2 BILLION LIGHT YEARS
ACROSS · ·
AS BIG AS A FIFTH OF
THE UNIVERSE THAT
WE CAN SEE SO FAR !

TO SOME,
EVIDENCE
OF THE
"BIG BANG"
IS THE
MICROWAVE
LIGHT
"BACKGROUND
RADIATION"
THROUGHOUT
THE UNIVERSE

SUPPOSEDLY,
THE REMAINING
ENERGY
OF THE "ORIGINAL"
HIGH ENERGY
PHOTONS
THAT WERE
TO HAVE
FILLED THE
BIG BANG
UNIVERSE
DURING THE
FIRST
HUNDREDS OF
THOUSANDS
OF YEARS

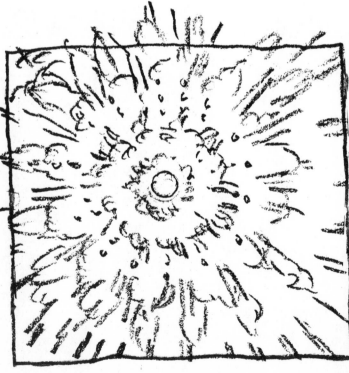

BUT,
THERE'S
LIGHT OF
OTHER
WAVELENGTHS

FLOATING
AROUND
UNSPOKEN
FOR · ·

AND
THERE ARE
WAYS
LIGHT
OF
MANY
WAVELENGTHS
COULD APPEAR
TODAY

LITTLE
WHITE HOLES
IN SPACE
?

QUASARS
?

PERHAPS · · ·

IF THE WATER AND ICE
CRYSTALS OF OUR OWN
ATMOSPHERE REFRACT
THE BLUE WAVELENGTHS
MORE THAN THE OTHERS
?

IF DUST SCATTERS
THE BLUE WAVELENGTHS
OF THE STARLIGHT
SPECTRUM
?

IF THE HALOS OF
THE GALAXIES REFRACT
THE BLUE WAVELENGTHS
OF THE LIGHT OF THE
STARS
?

PERHAPS THIS IS
AN ETERNAL UNIVERSE

FOREVER NEW
FOREVER OLD

JUST AS IT LOOKS
TO THE CHILDREN
OF THE WORLD

A GREAT SKY
WITH A LOT OF STARS..
VAST ..
AWE INSPIRING..
WONDERFUL AND RICH

THAT SEEMS TO
GO ON FOREVER

LIGHT

SOME SAY THERE
IS AN ETERNAL
"TIMELESS
LIGHT"
THAT ACTS
IN
PERFECT
RESPONSE
TO ALL THE
WORLD

HERE IS
AN ATTEMPT
TO "PICTURE"
THIS
TIMELESS
LIGHT . . .

LITTLE TWINKLES

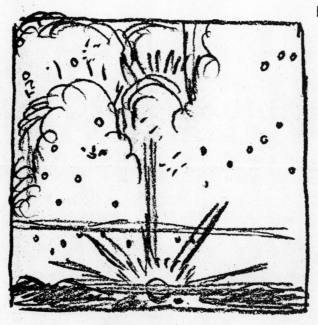

IN THE OCEAN OF LIGHT -
MATTER APPEARS
OUT OF A
KIND OF NOTHINGNESS
AT THE MEETINGS
OF THE WAVES

MATTER CONTINUALLY
FLASHES IN AND
FLASHES OUT . .
APPEARING AND
DISAPPEARING
WITH THE MOVEMENT
OF THE WAVES

IN THE LITTLEST VIEW,
EMPTY SPACE IS
A FOUNTAIN OF LIGHT
LIGHT APPEARS
WHERE "MATTER"
DISAPPEARS

ATOMS BEHAVE
IN TWO WAYS
AT THE SAME TIME...

LIKE PARTICLES
AND LIKE "WAVES"

ALL OF SPACE
CAN BE VIEWED AS
A "HARMONY" OF
WAVES OF
ATOMS

ATOMS
ARE
WAVE HARMONIES
OF
LIGHT

ALL OF SPACE IS AN OCEAN OF LIGHT

THE ATOM

WE ARE
MADE OF
ATOMS

IT WAS ONCE THOUGHT THAT THE ATOM WAS THE LITTLEST POSSIBLE UNIT OF MATTER ·· SOMETHING INDIVISIBLE AND FUNDAMENTAL

EARLY IN THE 20TH CENTURY, IT WAS DISCOVERED THAT THE ATOM WAS MADE OF TINIER PARTICLES.

IN 1911, THE ATOM CAME TO BE VIEWED AS A "CLOUD" OF ELECTRONS ·· WITH A TINY NUCLEUS AT THE CENTER

AND THEN THE NUCLEUS WAS FOUND TO BE MADE OF EVEN TINIER PARTICLES.

PROTONS AND NEUTRONS ··

AND THE PASSING OF THE YEARS BROUGHT MANY KINDS OF SUB·ATOMIC PARTICLES TO LIGHT·· THEMSELVES THOUGHT TO BE INDIVISIBLE ··

ELECTRONS WHIZZING AROUND

FORM THE REGIONAL CLOUD "SHELLS"

THE ATOM OF TODAY

IS VIEWED AS A "NUCLEUS" (MADE OF PROTONS AND NEUTRONS) SURROUNDED BY A "CLOUD" OF ORBITAL ELECTRONS

WHICH RIDE IN A NUMBER OF HIGH PROBABILITY REGIONS CALLED "QUANTIZED ORBITALS"

AN ATOM IS AN ELECTRON·PROTON·PHOTON SYSTEM··

WHEN AN ELECTRON IS IN THE NEAREST ORBIT TO THE NUCLEUS·· IT'S IN IT'S LOWEST ENERGY STATE

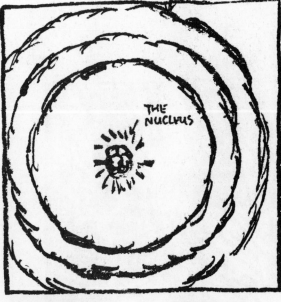

THE NUCLEUS

AN ATOM IN THE LOWEST ENERGY "GROUND" STATE ·· WILL STAY THERE FOREVER ·· UNLESS IT'S EXCITED BY CERTAIN WAVELENGTHS OF LIGHT ··

WHEREUPON IT WILL "QUANTUM JUMP" TO A HIGHER ORBIT REGION

IN AN "ATOMIC QUANTUM JUMP" AN ELECTRON MOVES IN A "SPACE LIKE PATH" BETWEEN TWO "TIME LIKE PATHS" AND APPEARS TO MOVE INSTANTANEOUSLY !

"AS AN ELECTRON QUANTUM JUMPS" TO A HIGHER ORBIT·· ENERGY STATE OF THE ATOM·· IT DEMATERIALIZES AND REMATERIALIZES, ABSORBING A PHOTON OF LIGHT OF AN EXACT COLOR

"AS AN ELECTRON QUANTUM JUMPS" TO A LOWER ORBIT·· ENERGY STATE OF THE ATOM·· IT DEMATERIALIZES AND REMATERIALIZES, EMITTING A PHOTON OF LIGHT OF AN EXACT COLOR

ABOUT 90% OF THE ATOMS OF THE UNIVERSE ARE THOUGHT TO BE ATOMS OF HYDROGEN ··
AN ATOM OF HYDROGEN IS A ONE PROTON NUCLEUS·· ORBITED BY ONE ELECTRON
THE NEXT MOST ABUNDANT ATOM IN THE UNIVERSE ARE ATOMS OF HELIUM ···
AN ATOM OF HELIUM IS A NUCLEUS MADE OF 2 PROTONS AND 2 NEUTRONS ORBITED BY 2 ELECTRONS

THE NUMBER OF PROTONS IN THE NUCLEUS IS CALLED THE ATOMIC NUMBER ··
IT DETERMINES THE NATURE OF THE ELEMENT AND THE NUMBER OF SITES AVAILABLE FOR ELECTRONS
FOR MANY YEARS THERE WERE THOUGHT TO BE 92 ELEMENTS EXISTING IN THEIR NATURAL STATE IN THE WORLD ·· WITH FROM 1 TO 92 PROTONS IN THE NUCLEUS AND VARYING NUMBERS OF NEUTRONS

IN RECENT TIMES, SOME "HEAVIER" ELEMENTS HAVE BEEN DISCOVERED IN NATURE
AND SOME HAVE BEEN SYNTHESIZED BY SCIENCE··
BUT THEY DO NOT LAST TOO LONG.

LIGHT MATTER

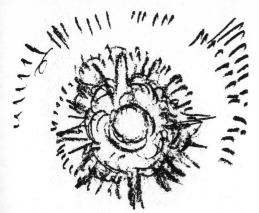  

AN ATOM
IS A
HARMONY
OF
WAVES
OF
"QUARKS"

NOW··
ALL MATTER
IS THOUGHT TO BE
MADE FROM TINIER
INDIVISIBLE, FUNDAMENTAL
PARTICLES··CALLED
"QUARKS"

THERE ARE "KINDS"
OF QUARKS··
THOSE THAT
MAKE UP THE
PROTONS AND
NEUTRONS ※'

ELECTRONS AND
NEUTRINOS ARE
THOUGHT TO BE
KINDS OF QUARKS

SOME SAY "QUARKS"
ARE TINY PHOTONS
ROTATING AS A RING

AS OF NOW··
A SOLITARY QUARK
HAS NOT BEEN ISOLATED
NOR SEEN, BUT THERE
IS SCIENTIFIC EVIDENCE
FOR ITS EXISTENCE

ALL THE LIGHT
THAT WE SEE
EITHER
EMENATES FROM
OR IS REFLECTED
FROM
ATOMS

AND
ATOMS THEMSELVES
ARE HARMONIES OF
WAVES OF LIGHT
IN AN OCEAN
OF LIGHT

PHOTONS
ARE THE SMALLEST
UNIT OF LIGHT··
OF ELECTROMAGNETIC
ENERGY

PHOTONS TRAVEL AT
THE SPEED OF LIGHT
AND CARRY A GREAT
RANGE OF ENERGY

LIGHT IS
A "STREAM"
OF
PHOTONS

THE CENTER
OF
EVERY
PHOTON
IS
·TIMELESS·

THE PROTON
IS THE FOUNDATION
OF ALL MATTER
IT HAS A POSITIVE CHARGE
AND IS THOUGHT TO BE
MADE OF
THREE "STRONG FORCE"
QUARKS

THE NEUTRON
IS A LITTLE HEAVIER
THAN THE PROTON··
AND ELECTRICALLY NEUTRAL
IT'S ALSO THOUGHT
TO BE MADE OF
THREE "STRONG FORCE"
QUARKS

THE ELECTRON
IS A FUNDAMENTAL
PARTICLE··
A "KIND" OF QUARK
IT HAS VERY LITTLE
MASS··
($1/1,836$TH OF A PROTON)
AND HAS A NEGATIVE
CHARGE

THE NEUTRINO IS ALSO
A FUNDAMENTAL PARTICLE··
A "KIND" OF QUARK··
IT HAS NO MASS
AND NO CHARGE··
AND IT TRAVELS AT
THE SPEED OF LIGHT

THE PHOTON··
IS A "PACKET"
OF ENERGY··
OF NO MASS
AND NO CHARGE··
IT TRAVELS AT
THE SPEED OF
LIGHT

MATTER FLASHES INTO EXISTENCE
FROM VIRTUAL PAIRS THAT ARE EVERYWHERE

ELECTRON/POSITRON
PAIRS CAN BE CREATED
BY HIGH INTENSITY
GAMMA RAYS

THE NEGATIVELY
CHARGED "ELECTRON"
IS VIEWED AS
MOVING FROM
THE PAST TO
THE FUTURE ...

THE POSITIVELY
CHARGED "POSITRON"
IS VIEWED AS
MOVING FROM
THE FUTURE
TO THE PAST ..
BACKWARDS
IN TIME !

THE ELECTRON AND
THE POSITRON
SPIN IN OPPOSITE
DIRECTIONS

FROM A
QUANTUM VIEW,
PAIRS OF
PARTICLES/
ANTIPARTICLES
ARE
CONTINUALLY
APPEARING
AND
DISAPPEARING
THROUGHOUT
ALL OF SPACE
! ! !

PARTICLES
ARE CREATED
IN PAIRS ..
PARTICLE
AND
ANTIPARTICLE

THE ELECTRON-
PHOTON-
POSITRON

THE
"FORWARD IN TIME"
AND
"BACKWARD IN TIME"
VERSIONS OF
THE SAME
THING

THE
PAIRS MOVE
APART ONLY
FOR A VERY
LITTLE
WHILE ..

AND THEN
COME BACK
TOGETHER
AND
DISAPPEAR
INTO EACH
OTHER

ANTI MATTER
IS THE ASPECT
OF MATTER
THAT TRAVELS
BACKWARDS
IN
TIME !

PROTON/ANTIPRO
CAN BE CREATED
BY HIGH INTENSITY
GAMMA RAYS

THE PROTON MOVE
FORWARD IN TIM

THE ANTI PROTON
MOVES BACKWAR
IN TIME !

A HIGH ENERGY
GAMMA RAY
CAN SPONTANEOU
CHANGE INTO A
PARTICLE/
ANTIPARTICLE

AS LONG AS IT IS
OF A GREATER E
THAN THE PARTIC
IT CREATES "

CLOUDS
OF
MANY
COLORS

LOOK ALIKES !
"BOUND STATES" OF
PARTICLE/ANTIPARTICLES

WAVE FUNCTIONS
OF "BOUND SYSTEMS"
OF "QUARKONIUM"
(A HEAVY QUARK/
ANTI QUARK
TOGETHER

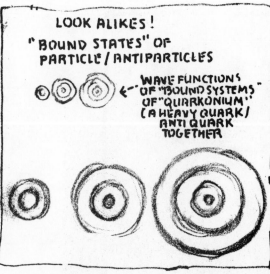

"QUARKONIU
IS 100,000
LITTLER T
"POSITRON

IS UP TO
10,000 TIM
MORE ENE
THAN
POSITRON

POSITRON
.. AN ELECT
AND A
POSITRO
TOGETH

THERE APPEAR TO BE
DIFFERENT "FORCES"
IN NATURE ..

· THE GRAVITATIONAL FORCE
· THE ELECTROMAGNETIC FORCE
· THE NUCLEAR FORCE

(SOME SAY ALL FORCES
ARE DIFFERENT EXPRESSIONS
OF THE GRAVITATIONAL FORCE)

THE "PARTICLES" OF
EACH FORCE ARE
CONNECTED BY
THEIR RESPECTIVE
"QUANTUM", OR
VIRTUAL,
WAVELIKE ASPECT

THE ELECTROMAGNETIC
FORCE

PROTON/ANTIPROTON
NEUTRON/ANTINEUTRON
ELECTRON/POSITRON ✳
QUARK / QUARK
NEUTRINO / ANTINEUTRINO
QUARK / QUARK
PAIRS
ARE CONNECTED BY THE
WAVELIKE ASPECT OF THE
ELECTROMAGNETIC FORCE ..
CALLED "..."PHOTONS"

THE NUCLEAR FORCE

THE TYPES OF QUARKS OF WHICH
PROTONS AND NEUTRONS ARE MA
ARE CONNECTED BY THE "NUCLEAR"
OR "STRONG" FORCE
THE STRONG FORCE ACTS AT VERY
SHORT RANGES ONLY ..
THE QUANTUM, OR WAVELIKE
ASPECT OF THE STRONG FORCE
ARE CALLED "GLUONS"

"QUARKONIUM" .. THE BOUND STATE
OF A "HEAVY QUARK / ANTIQUARK"
IS SAID BY SOME, TO BE HELD
TOGETHER BY THE "COLOR FORCE"
WHICH IS THOUGHT
TO BE THE FOUNDATION OF
ALL NUCLEAR FORCES

✳ THE "WEAK FORCE" ..
ACTING ONLY AT VERY
SHORT RANGES ..
IS CONSIDERED BY SOME
TO BE AN EXPRESSION
OF THE ELECTROMAGNETIC
FORCE

ELECTRON / POSITRON
QUARK / QUARKS
NEUTRINO / ANTINEUTRINO
QUARK / QUARKS

ARE CONNECTED BY THE
WAVELIKE ASPECTS OF
THE WEAK FORCE "BOSON

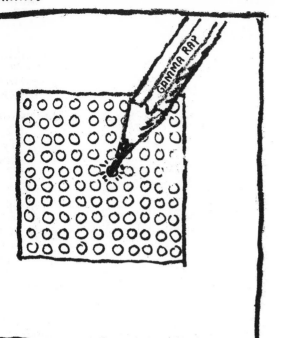

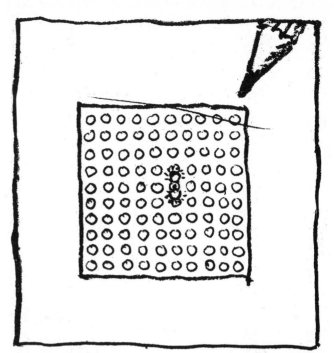

O UNDERSTAND
HE PROCESS OF
AIR CREATION,
LL OF EMPTY SPACE
S VIEWED AS BEING
ILLED WITH "VIRTUAL",
R "IMAGINARY"
AIRS OF PARTICLES..

IMAGINARY PROTONS
UPON IMAGINARY
ANTIPROTONS..
IMAGINARY ELECTRONS
UPON IMAGINARY
POSITRONS..
AND SO FORTH
AND SO ON..

WHEN AN IMAGINARY
PAIR OF PARTICLES
IS TOUCHED BY
A HIGH ENERGY
GAMMA RAY,
THEY'RE BROUGHT
INTO REALITY.

ABSORBING ENOUGH ENERGY
TO CHANGE INTO MATTER
IN ACCORDANCE WITH
$E = mc^2$
A PAIR CREATION..
MATTER AND ANTIMATTER
IN THE REAL WORLD!

HE GRAVITATIONAL FORCE
ONNECTS
LL PARTICLES
HROUGHOUT
LL OF SPACE-TIME

TS VERY WEAK
OMPARED TO THE
THER FORCES

GRAVITATIONAL WAVES
CARRY ENERGY FROM
EVENT TO EVENT
AT THE SPEED OF
LIGHT
AS THEY TRAVEL
FURTHER AND FURTHER...
THEY BECOME FAINTER
AND FAINTER

SOME SAY
GRAVITATIONAL
WAVES ARE
WAVES
OF
SPACE-TIME
ITSELF

..THAT
"MATTER"
IS
HIGH
"CONCENTRATIONS"
OF
SPACE-TIME
WAVES!

OMEDAY, PERHAPS, WE'LL
E ABLE TO LOOK TO THE STARS
N GRAVITY WAVE LIGHT..
ND NEUTRINO LIGHT
STRONOMY

THERE IS A RANGE
OF ENERGIES OF THE
GRAVITATIONAL SPECTRUM
JUST AS THERE IS A
RANGE OF ENERGIES
IN THE ELECTROMAGNETIC
SPECTRUM

WAVELIKE ASPECTS
OF THE
GRAVITATIONAL FORCE
ARE CALLED
"GRAVITONS"

GRAVITATIONAL WAVES
AT THE QUANTUM LEVEL
COULD BE THE..
"FOUNDATION"
OF ALL MATTER

HERE IS A VIEW
.. AT THE TINIEST LEVELS
.. OF ALL OF SPACE
AS AN OCEAN
OF LIGHT

LITTLE
TWINKLES

BUBBLES
ORIGINATING
FROM
THE PAST
AND
FROM THE
FUTURE

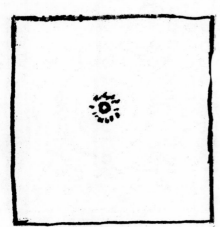

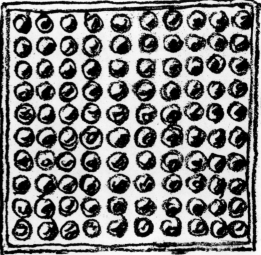

AT THE TINIEST LEVEL
ALL TIME EXISTS
AT THE SAME TIME ..
AT TINY SINGULARITIES *
THAT EXIST
EVERYWHERE

TINY BUBBLES
SURROUNDING
THE SINGULARITIES

TWINKLE
'ON' AND 'OFF'

IN THIS VIEW ..
ALL OF SPACE
IS AN OCEAN OF TINY
"VIRTUAL" BUBBLES
CONTINUALLY
APPEARING AND
DISAPPEARING

THE BUBBLES
ARE THE
BEGINNINGS
OF THE
SPACE-TIME
WAVES

SINGULARITIES
APPEAR TO
CREATE
ENERGY AND
LIGHT
FROM
"NOTHINGNESS"

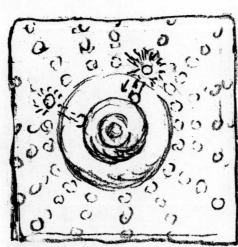

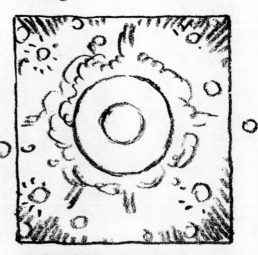

LIGHT
RADIATES
FROM
THE
SINGULARITY

PAIRS OF
PARTICLES
APPEAR AND
DISAPPEAR
IN ITS VICINITY

NEAR A
" LITTLE WHITEHOLE"
ONE MEMBER OF A PAIR
MAY FALL THROUGH
THE EVENT HORIZON ..
AND THE OTHER ONE
IS "FREE" AND IT
APPEARS !

MATTER AND ANTIMATTER
APPEAR TO FLOW FROM A
SINGULARITY

AS PARTICLES DISAPPEAR
NEW ONES ARE CREATED
IN THE INTENSE GRAVITY
NEAR IT.

---

* A SINGULARITY
IS BEYOND TIME ..
IT CAN'T BE DESCRIBED
IN TERMS OF THE
KNOWN SCIENCES
OF TODAY

THE OCEAN OF
VIRTUAL BUBBLES
IS CALLED
"THE QUANTUM FOAM"

THESE BUBBLES ARE THEORETICAL ..
THEY MAY OR MAY NOT EXIST ..
WE CAN'T REALLY SEE ANYTHING
THAT TINY ...

RIGHT NOW ITS JUST A WAY OF
TRYING TO UNDERSTAND
THE WORKINGS OF THE UNIVERSE
IN TERMS OF ACCEPTABLE THEORY

A
RING
SINGULARITY

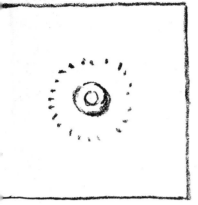

A WHITEHOLE··

A "RING SINGULARITY
WITHIN AN EVENT
HORIZON

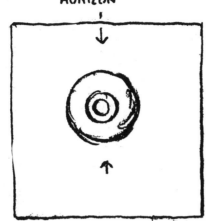

THE
LITTLE
WHITE
LIGHT

A ROTATING
PHOTON
A QUARK
A WHITEHOLE

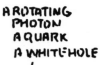

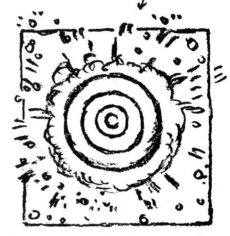

HE SINGULARITIES
APPEAR
IN
SPACE·TIME
AS
WHITEHOLES

A WHITEHOLE IS
A PHOTON THAT IS
ROTATING AS A
"RING SINGULARITY"
WITHIN AN
EVENT HORIZON

LIGHT RADIATES
FROM THE
SINGULARITY
OF THE
WHITEHOLE

THE LIGHT EMITTED
BY A WHITEHOLE
PASSES THROUGH
THE EVENT HORIZON
INTO THE UNIVERSE

JUST AS THE CENTER
OF EVERY PHOTON
IS TIMELESS··
THE CENTER OF
EVERY WHITEHOLE
IS TIMELESS !

INY WHITEHOLES
OME INTO BEING
UT OF A KIND OF
NOTHING·NESS"

PERHAPS THEY APPEAR
WHEREVER AND
WHENEVER
AS NEEDED ... TO
COSMICALLY BALANCE
THE UNIVERSE !

THE
WHITE·HOLE
LIGHT·HOLE ...

THE BEGINNING
AND
THE END
OF
LIGHT

FROM
AND
OF
TIMELESSNESS

ALL OF TIME

THE INSTANTANEOUS CONNECTIONS

REFLECTIONS
LIKE
"LANDSCAPES"
IN
RAINDROPS

A
WAVE
HARMONY
OF LIGHT

A FIELD IN MOTION
APPEARS TO
GENERATE MATTER
AS THE SINGULARITIES
DELIVER IT
"ON THE SPOT"

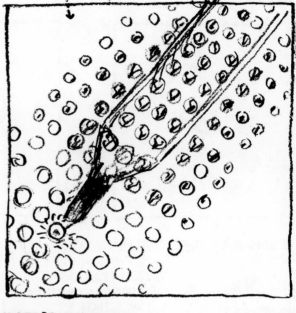

FROM SUPERSPACE
ALL TIME EXISTS
AT THE SAME TIME

HARMONIES OF WAVES
AT THE
"REAL"
"COSMIC"
AND
"HEAVENLY"
LEVELS OF
PERCEPTION

THE SINGULARITY
ACTS
IN
TIMELESSNESS
IN
ALL OF TIME

EMENATE
FROM A
QUIET TIMELESS
BEAUTY

ESTABLISHING
WAVE
HARMONIES
IN
REAL
TIME

PERHAPS THERE ARE
"FAMILIES" OF WAVES
SENT BY FAMILIES AND
FRIENDS THROUGHOUT
TIME!
WE "RIDE" THE WAVES
LIKE LIGHT ON THE
WATERS

WAVES FROM
LITTLE
TWINKLING
BUBBLES

JOIN INTO
WAVE
HARMONIES
OF
ATOMS

MATTER APPEARS
AT THE HARMONIES
OF THE
WAVES

CONNECTIONS

APPEARING
AND DISAPPEARING
"BRIDGES"
CONNECT EVERY PART
OF SPACE-TIME
WITH EVERY OTHER

IN A QUANTUM
JUMP
THE WAVE FUNCTION
APPEARS TO
COLLAPSE IN TIME

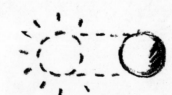

A "QUANTUM JUMP"
IS INSTANTANEOUS!...
SOMETHING DISAPPEARS
IN ONE PLACE AND
REAPPEARS IN ANOTHER
AT THE SAME INSTANT

QUANTUM JUMPS CAN HAPPEN
BETWEEN ANY POINTS
WITHIN THE LIGHT CONES...
OR WITHIN THE "ELSEWHERE"...
OR EVEN BETWEEN REGIONS
SEPARATED BY THE WALL
OF LIGHT!

AS THE QUANTUM WAVE
MOVES IN A "SPACE-LIKE"
FASTER-THAN-LIGHT PATH!

TIME
STOPS

TRAVELING
AT
THE SPEED
OF LIGHT

EVERYTHING
IS
TOTALLY
STILL

WAVES MOVING
AT SPEEDS LESS THAN
THE SPEED OF LIGHT...
MOVE IN
TIME-LIKE PATHS

"FASTER THAN LIGHT"
WAVES
MOVE IN
SPACE-LIKE PATHS

WAVES MOVING
AT THE SPEED OF LIGHT
MOVE IN
LIGHT-LIKE PATHS

THE LIGHT CONE

THE WALL OF LIGHT

THE WALL OF LIGHT
IS A "BUBBLE OF LIGHT"
IN ANOTHER
DIMENSION

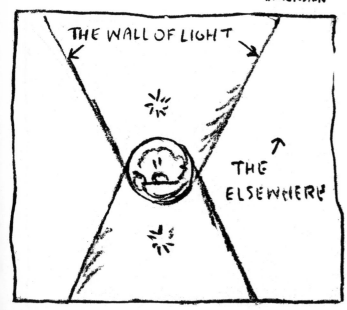

THE WALL OF LIGHT

THE ELSEWHERE

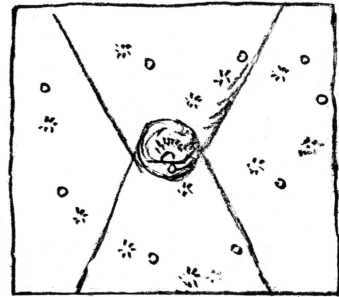

THERE IS, IN THE QUANTUM REALM, THE POTENTIAL FOR INSTANTANEOUS COMMUNICATION BETWEEN ALL EVENTS IN SPACE · · BOTH FORWARD AND BACKWARD IN TIME

REGARDING AN "EVENT AS A POINT IN SPACE-TIME EACH EVENT HAS A "FUTURE LIGHT CONE" CONTAINING ALL FUTURE CONSEQUENCES OF THE EVENT · · AND A "PAST LIGHT CONE" CONTAINING ALL PAST INFLUENCES UPON THE EVENT

ALL EVENTS WITHIN THE PAST AND FUTURE LIGHT CONES TRAVEL AT SPEEDS LESS THAN THE SPEED OF LIGHT · · THEY'RE SEPARATED FROM THE "ELSEWHERE" BY A WALL OF LIGHT

ALL EVENTS OUTSIDE THE LIGHT CONES ARE IN THE "ELSEWHERE" · · IN THE ELSEWHERE · · COMMUNICATION CAN BE INSTANTANEOUS ! FASTER THAN THE SPEED OF LIGHT · !!

EVENTS IN THE "ELSEWHERE" CAN INSTANTANEOUSLY INFLUENCE EVENTS WITHIN THE WALL OF LIGHT THROUGH THE FASTER THAN LIGHT FLOW OF QUANTUM WAVES

THE ONE THAT IS EVERYTHING

LIKE A GREAT SHIMMERING CRYSTAL

PHOTONS LIGHTING UP !

A TIMELESS PROCESS

A LIGHTER "REALITY" CAN APPEAR AT ANY TIME COMPLETE WITH ALL MEMORIES AND KNOWLEDGE

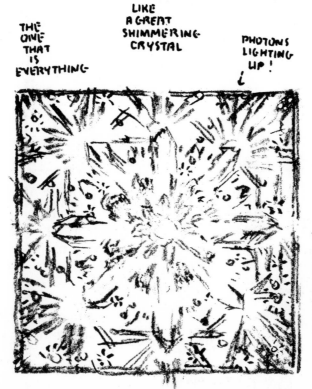

ALL THE PROBABLE FUTURES ALL THE PROBABLE PASTS ALL THE PROBABLE "NOW" MOMENTS ARE INSTANTANEOUSLY CONNECTED

THE TIMELESS BOUNDARY NEED NOT BE BILLIONS OF LIGHT YEARS AWAY AS "TIMELESSNESS" IS AT THE CENTER OF THE LIGHT WE ARE MADE OF

ALL TOGETHER NOW

THE QUALITY
OF LIFE
EXPERIENCE
TODAY
IS
DIRECTLY
AFFECTED
BY OUR
UNDERSTANDINGS
OF
HISTORY
AND
OUR EXPECTATIONS
FOR
THE
FUTURE

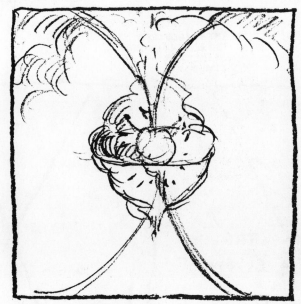

SOME SAY
THAT
"TODAY"
IS A LITTLE
MOVIE ...

A PERFECT
REFLECTION
OF
OUR VIEWS
OF
THE PAST
AND
THE FUTURE

THAT
THE PAST AND THE FUTURE
ARE
INSTANTANEOUSLY
UPDATED
REFLECTIONS
OF
THIS MOMENT
NOW

THE MOST PROBABLE
FUTURE
OF THE WORLD
IS CONTINUALLY
INSTANTANEOUSLY
UPDATED

IN ACCORDANCE
WITH OUR
VIEWS OF THE
PAST ..
AND OUR
HOPES FOR
THE FUTURE

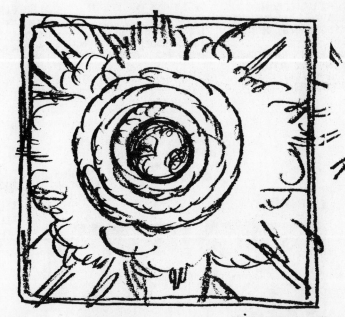

AT THE REAL WORLD
"COSMIC" AND
"HEAVENLY"
LEVELS OF
PERCEPTION

SOME SAY THERE IS
.. A KIND OF
"NOTHINGNESS"

THIS KIND OF "NOTHINGNESS"
IS CALLED A "WAVE FUNCTION"
IN THE LANGUAGE OF
THE SCIENCE OF TODAY

IT CARRIES
INSTANTANEOUSLY
UPDATED INFORMATION
OF WHAT MIGHT
HAVE HAPPENED ..
OF WHAT CAN
HAPPEN !

ITS SUBTLY AFFECTED
BY OUR EVERDAY
OBSERVATIONS.
IMPRESSIONS.
AND
EXPECTATIONS

AN INFINITE UNIVERSE WOULD
HAVE AN INFINITE NUMBER OF
STARS AND GALAXIES ..
AND WOULD, PHILOSOPHICALLY,
CONTAIN EVERY EVENT POSSIBLE
WITHIN THE LAWS OF NATURE *

FROM THE VIEW OF
"SUPERSPACE"..
OUR UNIVERSE
IS THEORETICALLY
CONNECTED TO AN
INFINITE NUMBER
OF PARALLEL,
COEXISTANT UNIVERSES!

WHICH COULD BE
INTERPRETED AS
BEING DIFFERENT
PAST AND FUTURE
VERSIONS OF THIS
REALITY ...

SUBTLY RELATED TO
THE IMPRESSIONS
AND EXPECTATIONS
OF THE OBSERVER

THE HARMONIES ARE
DIRECTLY RELATED TO
OUR IMPRESSIONS
OBSERVATIONS, AND
EXPECTATIONS ..
AT ALL LEVELS OF
PERCEPTION

THERE IS
WITHIN US ALL AND
WITHIN EVERY LITTLE THING

SOMETHING THAT CAN'T BE
DESCRIBED IN WORDS . .
SOMETHING THAT HOLDS
IT ALL TOGETHER . .

PLANET
EARTH
AS VIEWED
FROM
SPACE

THE "PAST"
IS A HARMONY OF
AN INFINITE VARIETY
OF HISTORIES . .
SOME MORE PROBABLE
THAN OTHERS . .

THE "FUTURE"
IS A HARMONY OF
AN INFINITE VARIETY
OF FUTURES . .
SOME MORE PROBABLE
THAN OTHERS . . .

"EVENTS" IN THE
WORLD OF TODAY
ARE ENACTMENTS
OF INSTANTANEOUS
UPDATINGS

FROM THE MOST
PROBABLE PAST
AND
FROM THE MOST
PROBABLE
FUTURE ! !

THERE IS
AN ETERNAL
EVERPRESENT
LIGHT

ACTING IN
PERFECT RESPONSE
TO THE
VIEWS
AND
HOPES
OF
ALL OF US

THROUGHOUT
ALL OF TIME

WHEN WE THINK ABOUT SOMETHING
THERE'S A LITTLE CHANGE OF
ENERGY
THAT CHANGES THE OBSERVER
AND THE OBSERVED

WHEN WE LOOK AT SOMETHING . .
THERE'S A LITTLE CHANGE OF
ENERGY
THAT CHANGES THE OBSERVER
AND THE OBSERVED

WONDER

THERE ARE THOUSANDS OF STORIES
OF HOW "ALL THIS" BEGAN...
STORIES AS DIFFERENT AS
THE PEOPLE AND PLACES AND
HISTORIES OF THE WORLD

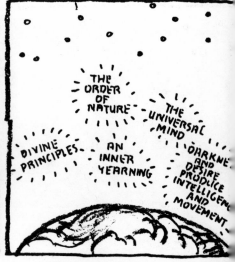

ANCIENT CIVILIZATIONS
THROUGHOUT TIME
HAVE HAD THEIR VIEWS✳
OF HOW THE "WORLD"
BEGAN OR BEGINS

RELIGIONS
THROUGHOUT TIME
HAVE HAD THEIR VIEWS✳
OF HOW THE "WORLD"
BEGAN

PHILOSOPHIES
THROUGHOUT TIME
HAVE HAD THEIR VIEWS✳
OF HOW A UNIVERSE
BEGINS

FROM HERE ON EARTH
ITS VERY DIFFICULT
TO KNOW FOR SURE
HOW OR WHY
THIS "UNIVERSE"
BEGAN..
OR IF IT HAD
A BEGINNING
AT ALL

PERHAPS..
THE VALUABLE
QUESTIONS
FOR THE WORLD OF
TODAY ARE:

WHAT IS THE BEGINNING
OF
LIGHT?
OF
THE STARS?
OF
THE
GALAXIES
?

WHERE DO WE COME FROM?

✳REFLECTIONS OF THEIR OWN
OBSERVATIONS, IMAGES,
INTERESTS, AND
UNDERSTANDINGS OF
HISTORY AND
INSPIRATIONS

EACH OF US
FINDS
THE TRUTH
IN
OUR
OWN
WAY

SCIENCES
THROUGHOUT TIME
HAVE HAD THEIR VIEWS☀
OF HOW THE "WORLD"
OR "UNIVERSE" BEGAN

AND THE CHILDREN OF THE WORLD
HAVE THEIR VIEWS⚹
TOO

NOW..

KNOWING WHAT WE KNOW TODAY
HOW CAN THE LIFE EXPERIENCE
BECOME A LITTLE LIGHTER
FOR EACH AND EVERY ONE OF
US
?

DO YOU EVER WONDER
WHERE WE COME FROM?

TINY ATOMS
OVER BILLIONS
OF YEARS...
COMING TOGETHER
BY CHANCE! (??)

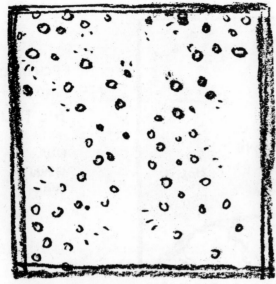

DO WE COME FROM
CHANCE ENCOUNTERS OF
TINY ATOMS?

DO WE COME FROM THE STARS?

IF THERE IS
A "BIG PICTURE"

WHY CAN'T WE
SEE
IT
?

SOME SAY
THE WORLD
IS
EXACTLY
AS IT APPEARS
NO MORE...
NO LESS...

SOME SAY
THE WORLD
IS
WITHIN
A
GREATER
EVERPRESENT
REALITY

ARE THERE OTHER REALITIES
EXISTING CONCURRENTLY
WITH OURS?

ARE THERE GUIDING FORCES?
ARE THERE GUIDING LIGHTS?
ARE THERE ANGELS?

IF THERE IS A "BIG PICTURE"
WHY CAN'T WE SEE IT?

ARE THE HEAVENS
THE RESULT OF OUR IMAGINATIONS
THROUGHOUT TIME?

OR ARE WE ACTUALLY
OF THE HEAVENS?

WHERE DO WE COME FROM?
WHERE DO WE GO TO?

IS LIFE
WHAT
WE
MAKE
IT
?

IS OUR
DESTINY
WRITTEN
IN THE
STARS
?

ARE WE DOING
WHAT WE'RE DOING
BECAUSE:

WE ARE
DOING WHAT'S RIGHT
FOR OURSELVES,
OUR FAMILY
OUR FRIENDS?

IS IT BECAUSE WE
ARE ACTING IN
PERFECT RESPONSE
TO THE YEARNINGS
OF THE WORLD?

HAVE WE BEEN GIVEN
A MISSION FROM
A HEAVENLY
TIME AND PLACE?

ARE WE COSMICALLY
DOING WHAT WE'RE DOING
FOR THE BEAUTY AND
KINDNESS IT LEAVES
IN ITS WAKE?

IS IT ALL HAPPENING
AT THE SAME TIME?

FOR THOSE WHO PREFER
A "REAL WORLD" VIEW OF LIFE

HERE IS A VIEW OF THE HEAVENS

INSPIRED BY THE
EVERDAY WORLD
OF TODAY

WE LEARN BY
SEEING THINGS IN PLAIN VIEW
AND
EXCHANGING IDEAS

HOW CAN
WE SEE
THE BEST
POSSIBLE
"PICTURE"
OF
OUR
"WORLD"
?

TODAY, WE CAN SEE
THE WONDER AND BEAUTY
OF A LOT OF THE
KNOWN UNIVERSE
IN FULL COLOR SLIDES

SOON IT'LL BE POSSIBLE
TO SHOW THE WORLD
THROUGHOUT HISTORY
AS A PROJECTED LIGHT MOVIE
IN FULL COLOR AND SOUND
AND VARIABLE SPEED

TO SHOW
FULL COLOR
PROJECTED LIGHT MOVIES
OF THE WORLD'S MOST POPULAR
STORIES OF
COSMIC BEGINNINGS

THE CHILDREN OF THE WORLD
AND THE EVERY DAY PEOPLE
OF THE WORLD
DESERVE A CHANCE TO
TELL THEIR IDEAS OF
WHAT THE HEAVENS LOOK LIKE

WE CAN DO THIS IN
THE QUIET AND BEAUTY
OF THE GREAT OUTDOORS
IN THE TRADITION OF
THOUSANDS OF YEARS

IT'D BE WONDERFUL TO SEE
WHAT THESE IDEAS LOOK LIKE
IN "LIGHT"
FROM OUR HEARTS

A TALE OF TALES
OF WHERE WE
MIGHT HAVE COME
FROM
OF WHERE WE MIGHT GO TO
SOME DAY

ZOOMING INTO
A NEARBY
STARCLOUD

ZOOMING INTO
A GALAXY
FARAWAY

...STARS BEGINNING
AND GALAXIES BEGINNING
AND LIGHT ITSELF BEGINNING

RIGHT IN FRONT OF OUR EYES!

AND SHOW
OUR IDEAS
OF WHAT THE HEAVENS
MIGHT LOOK LIKE

IN GLORIOUS COLOR
WITH THE MOST POPULAR MUSIC
OF OUR TIME

THE FIRST STAR SHOWS
WOULD PROVIDE THE
EVERYDAY PEOPLE
AND THE CHILDREN OF THE WORLD
THE THRILL OF APPEARING
ANYWHERE
IN THE "HEAVENS" OF ALL TIME

WE COULD IN A LIGHT HEARTED
WAY
BRING "THE HEAVENS" TO EARTH

AND GET A PICTURE
OF
WHERE WE COME FROM

FOR THOSE WHO PREFER
A COSMIC VIEW OF LIFE

HERE IS A VIEW OF A
"TIMELESS" COSMIC HEAVEN

NOW AND FOREVER,
WE COME FROM WHERE ITS SO
BEAUTIFUL AND PEACEFUL
THAT NO PLACE COULD POSSIBLY
BE ANY LOVELIER OR ANY BETTER

WE COME FROM
A LAND OF TIMELESS BEAUTY AND GRACE

DIMMING THE GLORY
FROM THE MEMORIES OF OUR MINDS
BUT NOT FROM OUR HEARTS

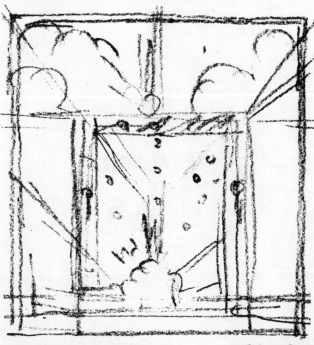

ALL OF TIME
IN
QUIET BEAUTY

IN TIMELESS BEAUTY
WE PROJECT OUR LIGHT
THROUGHOUT ALL OF TIME

WE EXIST
IN TIME
AND IN TIMELESSNESS
AT THE SAME TIME.

WE SOMETIMES GET
GLIMPSES OF THE FUTURE
AND OF THE PAST
BECAUSE
FROM THIS COSMIC VIEW
WE EXIST NOW
IN THE PAST, PRESENT AND FUTURE

LITTLE MOMENTS
THROUGHOUT ALL OF TIME
LEAD US TO THE RIGHT PLACE
AND TIME
AND STATE OF BEING
FOR THE RETURN HOME

FROM THIS VIEW,
LIFE IS A BRIEF MOMENT
AN ADVENTURE ..
A MYSTERY IF YOU PREFER ..
A PLAY IN WHICH WE
ARE ACTORS AND
PRODUCERS ..

HIDDEN FROM VIEW
AND UNKNOWN TO US
UNTIL THE TIME IS
RIGHT ..
A MOVIE "WITH A
HAPPY ENDING !

FOR A LITTLE COMIC RELIEF
WE CAN SEE THE WORLD
AS A GREAT ADVENTURE ..
MYSTERY .. LIGHT MUSICAL ..
SCIENCE FICTION .. EDUCATIONAL ..
TEARJERKER .. BLOCKBUSTER
MOVIE .. WITH A
GLORIOUS ENDING !

WE BROADCAST CLUES
FROM HOME
TO HELP US FIND OUR
WAY ..
RECEIVING THEM EVERY
NOW AND THEN AS
INTUITIVE FLASHES ..
OR HUNCHES

WHEN WE MEET SOMEONE
WE INSTANTLY LIKE .
PERHAPS WE COME FROM
THE SAME "PLACE" ..
AND KNOW EACH OTHER
IN TIMELESSNESS AND
IN TIME !

THE HERE AND NOW IS
AN EXPERIENCE THAT
IS BUT A MOMENT
THAT SEEMS LIKE A LIFETIME

LITTLE MIRACLES-
HAPPEN THROUGHOUT TIME
JUST AT THE RIGHT TIME

IN THE HEAVENS OF ALL TIME

THE MOST BEAUTIFUL MOMENTS
ARE HAPPENING NOW

A LITTLE LIGHT
APPEARS
WHEREVER
AND
WHENEVER
JUST
AT THE
RIGHT TIME

SOMEHOW
SOMEWHERE
SOMEDAY

WE ALL GO HOME AGAIN

FOR THOSE WHO PREFER
A HEAVENLY VIEW OF LIFE

HERE IS A VIEW OF HEAVEN
FROM SOME OF
THE POPULAR STORIES
OF THE WORLD OF TODAY

AN
ENTRANCE
TO HEAVEN

THERE ARE THOUSANDS OF STORIES
OF HEAVENS...
STORIES AS DIFFERENT AS THE
PEOPLE AND PLACES AND HISTORIES
OF THE WORLD

SOME TELL OF HEAVENS
IN THE SKY

SOME HAVE "ENTRANCES"
AND WE GO THERE
SOMETIME IN THE
FUTURE

ARE WE GOING TO HEAVEN?

DO WE COME FROM HEAVEN?

IS HEAVEN WITHIN US ALL NOW?

ANY MOMENT ON EARTH
CAN BE
A KIND OF
"HEAVEN"

WHEN WE ARE WITH LOVED ONES

GOOD HEALTH

GOOD WEATHER

GOOD FRIENDS

A GOOD JOB

ENOUGH MONEY
TO LIVE, WORK, AND REST
IN PEACE

WE ARE HEAVENLY
IN A QUIET WAY

EACH TIME WE ARE
KIND

EACH TIME WE CARE
ABOUT OTHERS

EACH TIME WE DO
THE RIGHT THING

AND THE WORLD IS
A LITTLE BETTER
BECAUSE OF IT

SOME SAY
"HEAVEN"
IS
WITHIN
US
ALL

KEEP THE FAITH !
DON'T GIVE UP HOPE !
LOVE IS ETERNAL !

SOME SAY
HEAVEN CAN APPEAR
ALL OVER THE WORLD TODAY
FOR EVERYONE AND ANYONE
WHO IS READY

HEAVEN IS WITHIN
EVERY ONE OF US
AND ALL THINGS

IT CAN APPEAR
ANY TIME
ANY WHERE
THROUGHOUT
ALL OF TIME

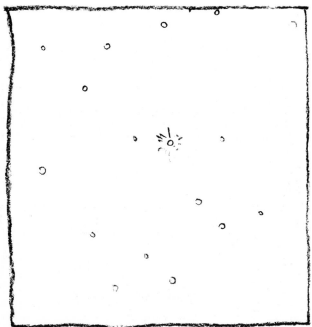

SOME SAY OUR DESTINY
IS "WRITTEN IN THE STARS"

SOME SAY OUR LIFE IS
WHAT WE MAKE IT

MAYBE
ITS A LITTLE
OF
BOTH

FOR THOSE WHO PREFER A

LIGHTHEARTED VIEW
OF LIFE

THROUGHOUT
ALL OF TIME . .

WE HAVE BROUGHT

THE WORLD

INTO
THIS
PERFECT
MOMENT

IT WILL ALL
HAVE BEEN
WORTHWHILE

HAVING
PASSED THROUGH
A GREAT STORM
IN THE
OCEAN OF TIME

THE WORLD
NOW
FLOATS FREE
AND CLEAR

HEAVENLY
ANGELS
SHOW
SOME FOLKS
THE WAY
IN A BEAUTIFUL
CRYSTAL
LIGHT

A SWEET
CHARIOT
SWINGS LOW
TO CARRY
SOME
FOLKS
HOME

ALL OVER THE WORLD
        EVERYBODY'S HAPPY ...   EVERYBODY FINDS JUST WHAT THEY'RE LOOKING FOR

ALL OVER
THE WORLD

THE HEAVENS
COME INTO VIEW

FOR SOME FOLKS,
GUIDING LIGHTS
APPEAR
IN THE SKY

FOR SOME,
THE WEARY ROADS OF LIFE
BECOME PATHS OF STARS

WHERE THE HEART IS TRUE

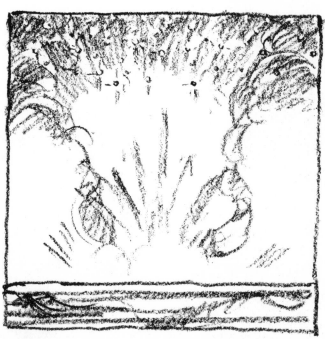

ALL OVER THE WORLD,
ALL PATHS LEAD
HOME

THE TREASURE OF TREASURES
THE DREAM OF DREAMS
THE HEAVEN OF HEAVENS
OUR HOME IN THE STARS

THE DREAM OF DREAMS

FOR THOSE WHO PREFER
A VERY QUIET AND GENTLE APPROACH

HERE IS
THE DREAM OF DREAMS

THE HEAVEN
THAT IS WITHIN
US ALL NOW
HAS BEGUN
TO MAKE
ITSELF
KNOWN

WE CHANGE THE WORLD OF TODAY
IN VERY QUIET WAYS
RIGHT NOW
THE MOMENT WE TRUST IN
OUR HOPES AND DREAMS
FOR A BEAUTIFUL TOMORROW

WITHIN OUR TIME
THIS WORLD WILL BE IN
LIGHT AND BEAUTY AND TRUTH
IT HAS BEGUN
IN VERY QUIET AND GENTLE WAYS

IN THIS DREAM OF DREAMS · ·
EVERYONE : . . IN TIME . . .

EACH IN OUR OWN WAY . . .
ARRIVES AT
A TRUE AND LASTING INNER PEACE

EVERYTHING IS JUST ABOUT THE SAME

BUT EVERYTHING LOOKS AND FEELS
A LITTLE LIGHTER

IT SEEMS LIKE A PARADISE
ITS ABSOLUTELY AMAZING
A LITTLE LIKE HEAVEN ON EARTH

WISHFUL THINKING

EVERYONE IS KIND

IT'S AS THOUGH
IT'S ALWAYS BEEN HERE
IT'S REALLY SERENE
THERE'S A QUIET IN THE AIR
THAT'S HARD TO DESCRIBE
IT'S SO BEAUTIFUL
IT MUST BE TRUE

THE WHOLE
WORLD
LIGHTENS
UP

IN THIS DREAM OF DREAMS
THE WHOLE WORLD
LIGHTENS UP...

A LOT SOONER THAN ANYONE
EVER THOUGHT POSSIBLE

IT'S SOMETHING LIKE
THE MOST WONDERFUL
DREAM OF DREAMS

IT'S SO BEAUTIFUL
IT MUST BE TRUE

ALL OVER THE WORLD
THE SKIES BEGIN
TO LIGHT IN
AMAZINGLY
BEAUTIFUL COLORS

IN THE BEGINNING
THE SKIES ARE A LITTLE LIGHTER
THE COLORS ARE AMAZINGLY
LOVELY

ALL OVER THE WORLD

EVERYTHING IS A LITTLE LIGHTER

THE BIBLIOGRAPHY

GENERAL READING FOR AN OVERVIEW

BOOKS

NATIONAL GEOGRAPHIC PICTURE ATLAS OF
OUR UNIVERSE
ROY A. GALLANT
NATIONAL GEOGRAPHIC SOCIETY 1980

THE AMAZING UNIVERSE
HERBERT FRIEDMAN
NATIONAL GEOGRAPHIC SOCIETY 1975

COLORPEDIA
THE UNIVERSE
THE RANDOM HOUSE ENCYCLOPEDIA
RANDOM HOUSE 1977

THE UNIVERSE
LIFE NATURE LIBRARY

LAROUSSE GUIDE TO ASTRONOMY
DAVID BAKER
LA ROUSSE 1979

THE BEAUTY OF THE UNIVERSE
HANS ROHR
TRANSLATED AND REVISED BY ARTHUR BAER
THE VIKING PRESS 1972

ATLAS OF THE UNIVERSE
PATRICK MOORE
RAND McNALLY 1970

CATALOG / SLIDES AND PRINTS
KITT PEAK NATIONAL OBSERVATORY
CERRO TOLOLO INTERAMERICAN OBSERVATORY
SACRAMENTO PEAK OBSERVATORY 1979

THE 1980 HAMMOND ALMANAC
STARS / PLANETS / SPACE
HAMMOND ALMANAC 1980

THE CAMBRIDGE ENCYCLOPEDIA OF
ASTRONOMY
EDITED BY SIMON MITTON
CROWN 1977

ASTRONOMY /
STRUCTURE OF THE UNIVERSE
WILLIAM J. KAUFMANN
MACMILLAN 1977

THE UNIVERSE
FROM FLAT EARTH TO QUASAR
ISAAC ASIMOV
AVON / DISCUS 1968

THE UNIVERSE
DONALD GOLDSMITH
W.A. BENJAMIN 1976

GALAXIES
TIMOTHY FERRIS
SIERRA CLUB BOOKS 1979

ARTICLES

THE INCREDIBLE UNIVERSE
KENNETH F. WEAVER
NATIONAL GEOGRAPHIC MAY 1974

ASIMOV'S GUIDE TO THE UNIVERSE
ISAAC ASIMOV
MYSTERIES OF THE COSMOS
NEWSWEEK / FOCUS JUNE 1980

HEAVEN AND EARTH
LIFE DECEMBER 1979

THE DISCOVERY OF OUR GALAXY
CHARLES A. WHITNEY
ALFRED A. KNOPF 1971

LA ROUSSE ENCYCLOPEDIA OF MYTHOLOGY
PAUL HAMLIN 1975

A DICTIONARY OF WORLD MYTHOLOGY
ARTHUR COTTEREL
G.P. PUTNAMS SONS 1980

THE HOLY SCRIPTURES

ASTRONOMY / A POPULAR HISTORY
DORSCHNER, FREIDEMANN, MARK & PFAU
VAN NOSTRAND REINHOLD 1975

COLLIERS ENCYCLOPEDIA
ASTRONOMY SECTION
MACMILLAN 1977

ENCYCLOPEDIA AMERICANA
ASTRONOMY SECTION
AMERICANA 1978

THE FABRIC OF THE HEAVENS
STEPHEN EDELSTON TOULMIN
HARPER 1962

COLORPEDIA
THE EARTH, LIFE ON EARTH,
HISTORY AND CULTURE,
MAN AND SCIENCE
THE RANDOM HOUSE ENCYCLOPEDIA
RANDOM HOUSE 1977

THE FAMILY OF MAN
EDWARD STEICHEN
PROLOGUE BY CARL SANDBURG
THE MUSEUM OF MODERN ART 1955

MOVIES

NOVA

CONNECTIONS

THE ASCENT OF MAN

THE UNIVERSE AND DR. EINSTEIN

COSMOS

UNIVERSE

THE WILD WORLD OF ANIMALS

AMERICA

LIFE ON EARTH

HEAVENLY LIGHTS

HUGE NEW TELESCOPE ARRAY
EXTENDS MAN'S CELESTIAL VISION
JOHN NEARY
SMITHSONIAN JULY 1978

WISE MEN FROM THE SOUTH
PEER TOWARD WHAT MAY BE
THE LIMITS OF THE UNIVERSE
STERLING SEAGRAVE
SMITHSONIAN APRIL 1977

THE NEW ERA IN ULTRA VIOLET ASTRONOMY
FRED ESPENEK
ASTRONOMY OCTOBER 1978

THE X-RAY UNIVERSE
THOMAS MARKERT & THOMAS L. FENCEK
ASTRONOMY JULY 1977

X-RAY STUDIES IDENTIFY INTERGALACTIC CLOUDS
ASTRONOMY MAY 1978

GAMMA RAY LASER IN THE SKY
SCIENCE NEWS DECEMBER 18 1976

GAMMA RAY BURSTS FROM DIFFERENT SOURCES
SCIENCE NEWS DECEMBER 18, 1976

PIONEER VENUS DETECTS GAMMA RAY BURSTS
ASTRONOMY SEPTEMBER 1978

COMPUTER ENHANCES PHOTOS
ASTRONOMY FEBRUARY 1977

NEW COMPUTER SIMULATES COSMIC EVENTS
ASTRONOMY NOVEMBER 1978

PHYSICIST DISCOVERS GALAXY
CIRCLED BY STARS
ASTRONOMY FEBRUARY 1978

A STARRY HALO AROUND A GALAXY
SCIENCE NEWS

EXPLORING THE UNIVERSE
ANDREW FRAKNOI
CHICAGO SUN TIMES MARCH 10, 1979

SOMETHING VERY STRANGE AND AWESOME
RONALD KOTULAK
CHICAGO TRIBUNE JULY 15, 1979

SPACE TELESCOPE MAY OPEN DOOR TO UNIVERSE
RONALD KOTULAK
CHICAGO TRIBUNE NOVEMBER 9, 1979

NEW PLANETS TO MAKE SEEN?
CHICAGO SUN-TIMES MARCH 31, 1981

ORBITING TELESCOPE SHEDS NEW LIGHT ON STAR
PATRICK YOUNG
CHICAGO SUN-TIMES JUNE 13, 1981

GREAT GALAXIES:
STARGAZING MADE PICTURE PERFECT
PETER GORNER
CHICAGO TRIBUNE DECEMBER 4, 1980

ASTRONOMERS FIND 3 DISTANT GALAXIES
CHICAGO TRIBUNE JUNE , 1981

ARS

ERNOVAE
NNE HOPKINS
RONOMY APRIL 1977

ERNOVAS
TRICK E. THOMSON
NCE NEWS JANUARY 29, 1977

RONOMY /
ERNOVAS AS STAR FORMATION TRIGGERS
NCE NEWS

A SUPERNOVA TRIGGER
FORMATION OF THE SOLAR SYSTEM?
NTIFIC AMERICAN OCTOBER 1978

EARCHERS PROPOSE BRAND X SUPERNOVA
RONOMY MAY 1978

UD COLLAPSE AND STAR FORMATION..
O HOW TO TRIGGER IT
RONOMY SECTION
NCE NEWS FEBRUARY 5, 1977

RS...WHERE LIFE BEGINS
E. DECEMBER 27, 1976

LLAR EVOLUTION
HAEL SEEDS
RONOMY FEBRUARY 1979

NETARY NEBULAE, PART I ...
N STARS LOSE THEIR COOL
ID DARLING
RONOMY MARCH 1979

AT MAKES NOVAE BLOW UP?
ORAH BIRD & JOSEPH PATTERSON
RONOMY JULY 1977

ICAL SUPERNOVA REMNANT
NTIFIED IN GUM NEBULA
RONOMY MAY 1977

D DWARF STARS
L JOHNSON
RONOMY JULY 1978

OWN DWARFS AND BLACK HOLES
C. TARTER
RONOMY APRIL 1978

OLUTION IN BINARY SYSTEMS
LLIAM K. HARTMANN
TRONOMY SEPTEMBER 1977

RS WITH COMPANIONS ..
RE THAN MEETS THE EYE
LLIAM K. HARTMANN
TRONOMY SEPTEMBER 1977

OUBLE STARS EXPLAIN PULSATING STARS
TRONOMY FEBRUARY 1977

RE NEUTRON STARS SHINY?
TRONOMY NOVEMBER 1978

LSARS NUMBERS DOUBLE
TRONOMY SEPTEMBER 1978

RAY STARS IN GLOBULAR CLUSTERS
ORGE W. CLARK
IENTIFIC AMERICAN OCTOBER 1977

RAY TELESCOPE SENDS FIRST PHOTO
F PROBABLE BLACK HOLE
HICAGO TRIBUNE OCTOBER 1978

OSMIC BLOWTORCH
EW YORK TIMES APRIL 6, 1978

OBSERVATION OF STAR MOMENTS AFTER BIRTH
NEW YORK TIMES JANUARY 7, 1978

DEATH OF STAR / BIRTH TO NEW ONES
NEW YORK TIMES MARCH 14, 1978

GAMMA RAYS AND THE ORIGIN OF
COSMIC RADIATION
TREVOR WEEKES
ASTRONOMY JUNE 1977

WHAT ARE GAMMA RAYS?
GERRIT L. VERSCHUUR
ASTRONOMY JUNE 1977

INTERPLANETARY PARTICLES AND FIELDS
JAMES A. VAN ALLEN
SCIENTIFIC AMERICAN SEPTEMBER 1975

GRAVITY, DUST, AND SOLAR NEUTRINOS
JOHN GRIBBIN
ASTRONOMY JUNE 1978

THE STRUCTURE OF INTERSTELLAR MEDIUM
CARL HEILES
SCIENTIFIC AMERICAN JANUARY 1978

SPECTRAL LINES INDICATE
UNIDENTIFIED MOLECULES
ASTRONOMY FEBRUARY 1977

INTERGALACTIC GAS:
TOWARD A CLOSED UNIVERSE
SCIENCE NEWS JULY 16, 1977

SWARMS OF STARS:
COSMIC CALIBRATORS
JANET ROUNTREE LESH
ASTRONOMY MARCH 1978

THE RED AND THE BLUE
SCIENCE AND THE CITIZEN
SCIENTIFIC AMERICAN NOVEMBER 1979

ASTRONOMICAL SPECTROSCOPY
ROBERT STENCEL, WILLIAM BLAIR
& SUSAN CONAT-STENCEL
ASTRONOMY JUNE 1978

BETWEEN STARS AND SPACE
ROBERT E. STENCEL
ASTRONOMY SEPTEMBER 1977

THE GALAXY'S BIGGEST BUBBLE
SCIENCE NEWS MARCH 5, 1977

LIFETIDES
LYALL WATSON
OMNI NOVEMBER 1978

SCIENTISTS PROPOSE INTERSTELLAR CLOUDS
AS POSSIBLE SITES FOR LIFE
ASTRONOMY AUGUST 1977

LIFE CLOUD
FRED HOYLE & CHANDRA WICKRAMASINGHE
OMNI FEBRUARY 1979

CAN LIFE EVOLVE IN ELLIPTICAL GALAXIES?
JOHN GRIBBIN
ASTRONOMY MAY 1979

RECONSTRUCTING THE EVOLUTION OF LIFE
J. HACK
TECHNICAL REVIEW MAY 1978

THE ORIGIN AND EVOLUTION OF
THE SOLAR SYSTEM
A. G. W. CAMERON
SCIENTIFIC AMERICAN SEPTEMBER 1975

THE SOLAR SYSTEM
CARL SAGAN
SCIENTIFIC AMERICAN SEPTEMBER 1975

URANUS AND NEPTUNE
MICHAEL J.S. BELTON
ASTRONOMY FEBRUARY 1977

WHY DO PLANETS HAVE RINGS?
LEE ANNE WILSON
ASTRONOMY DECEMBER 1977

PROBING THE PLANETS
NEWSWEEK SEPTEMBER 10, 1979

THE SUN
E. N. PARKER
SCIENTIFIC AMERICAN SEPTEMBER 1975

OUR SUN
JAY M. PASACHOFF
ASTRONOMY JANUARY 1978

PROBING THE NEAREST STAR
JAMES E. OBERG
ASTRONOMY AUGUST 1977

BEYOND CENTAURI
THOMAS R. SCHROEDER
ASTRONOMY APRIL 1978

THE SHAKY MACHINE
TRUDY E. BELL
ASTRONOMY FEBRUARY 1978

SEISMIC STUDIES OF SOLAR PULSATIONS
EXAMINE STRUCTURE OF SUN'S INTERIOR
ASTRONOMY MAY 1977

SOLAR HOLES ALTER MAGNETIC FIELD
ASTRONOMY AUGUST 1977

SUN SPINNING FASTER
ACCORDING TO SCIENTIST
ASTRONOMY APRIL 1977

THE FAINT YOUNG SUN AND WARM EARTH
SCIENCE NEWS MARCH , 1977

THE EARLY HISTORY OF PLANET EARTH
WILLIAM K. HARTMANN
ASTRONOMY AUGUST 1978

PAST ASTEROID BOMBARDMENT
MAY HAVE AIDED EVOLUTION
ASTRONOMY NOVEMBER 1977

THE EARTH
RAYMOND SIEVER
SCIENTIFIC AMERICAN SEPTEMBER 1975

WORLD OF PLANTS
DAVID BELLAMY
THE LIVING EARTH SERIES
THE DANBURY PRESS, GROLIER 1975

CLOUDS OF THE WORLD
RICHARD SCORER
STACKPOLE 1972

SNOW CRYSTALS
W. A. BENTLEY & W. J. HUMPHREYS
DOVER 1962

THE GALAXIES

GALAXIES
TIMOTHY FERRIS
SIERRA CLUB BOOKS 1979

BLACKHOLES, QUASARS AND THE UNIVERSE
HARRY L. SHIPMAN
HOUGHTON MIFFLIN 1976

GALAXIES, NUCLEI AND QUASARS
FRED HOYLE
HARPER AND ROW, 1965

THE REDSHIFT CONTROVERSY
FRONTIERS IN PHYSICS
GEORGE B. FIELD, HALTON ARP & JOHN BAHCALL
W. A. BENJAMIN 1973

THE GREAT SPIRAL: OUR MILKY WAY
ASTRONOMY SEPTEMBER 1978

MILKY WAY AND BEYOND
TIMOTHY FERRIS
SCIENCE DIGEST MARCH 1981

THE ORIGIN OF GALAXIES
MARTIN J. REES & JOSEPH SILK
SCIENTIFIC AMERICAN
COSMOLOGY   JUNE 1978

PRIMEVAL GALAXIES
DAVID L. MATER & RASHID A. SUNYAEV
SCIENTIFIC AMERICAN NOVEMBER 1979

EVOLVING QUESTIONS ABOUT GALAXIES
SCIENTIFIC NEWS   NOVEMBER 12, 1977

ARE GALAXIES HERE TO STAY ?
VIRGINIA TRIMBLE & MARTIN REES
ASTRONOMY JULY 1978

AN AMAZING NEW GALAXY IS FOUND
MARK CHARTRAND III & TRUDY E. BELL
SCIENCE DIGEST   WINTER 1979

OPTICAL EMISSION FOUND IN
THREE RADIO GALAXIES
ASTRONOMY MARCH 1978

RADIO GALAXY POTENTIAL LARGEST
RADIO SOURCE
ASTRONOMY FEBRUARY 1977

MODEL FOR DOUBLE SOURCE
RADIO GALAXIES
ASTRONOMY APRIL 1977

COSMIC JETS
ROGER D. BLANDFORD,
MITCHELL C. BEGELMAN & MARTIN J. REES
SCIENTIFIC AMERICAN MAY 1982

SURVEY OF GALAXIES SHEDS LIGHT
ON SHAPE
SCIENCE NEWS   FEBRUARY 5, 1977

SEQUENCE LINKS QUASARS,
RADIO, NORMAL GALAXIES
ASTRONOMY OCTOBER 1977

POSSIBLE LINK NOTED BETWEEN GALAXIES
AND BARRED SPIRALS
ASTRONOMY NOVEMBER 1977

A GALAXY AROUND BL LACERTAE
PHYSICAL SCIENCES
SCIENCE NEWS FEBRUARY 18, 1978

NGC-1199
HALTON M. ARP
ASTRONOMY SEPTEMBER 1978

THE CLUSTERING OF GALAXIES
GROTH, PEEBLES, SELDNER & SONEIRA
SCIENTIFIC AMERICAN NOVEMBER 1977

GALAXY CLUSTERS EXAMINED
ASTRONOMY   MARCH 1978

RICH CLUSTERS OF GALAXIES
WALLACE H. TUCKER
ASTRONOMY OCTOBER 1977

RICH CLUSTERS OF GALAXIES
PAUL GORENSTEIN & WALLACE TUCKER
SCIENTIFIC AMERICAN NOVEMBER 1978

THE CENTER OF OUR GALAXY
R. H. SANDERS & G. T. WRIXON
SCIENTIFIC AMERICAN APRIL 1974

GALACTIC CENTER: A MATTER OF ANTIMATTER
SCIENCE NEWS MAY 6, 1978

DOES GRAVITY WAVE ?
DIETRICK E. THOMSEN
SCIENCE NEWS MARCH 18, 1978

SOLID SUPPORT FOR GRAVITY'S MIRAGE
SCIENCE NEWS   NOVEMBER 10, 1979

RESEARCHERS DETECT GRAVITY WAVES
ASTRONOMY FEBRUARY 1979

THOSE BAFFLING BLACK HOLES
TIME   SEPTEMBER 4, 1978

IS MASSIVE BLACKHOLE AT CENTER
OF MILKY WAY ?
ASTRONOMY NOVEMBER 1977

A HOLE IN THE MIDDLE OF THE GALAXY
DIETRICK THOMSEN
SCIENCE NEWS FEBRUARY 19, 1977

BLACK HOLE THEORY WEAKENS
ASTRONOMY SEPTEMBER 1978

ARE BLACK HOLES REALLY THERE ?
GREGORY A. SHIELDS
ASTRONOMY OCTOBER 1978

THE SEARCH FOR BLACK HOLES
KIP S. THORNE
SCIENTIFIC AMERICAN
COSMOLOGY   DECEMBER 1974

A JET BLACK HOLE IN A RADIO GALAXY
SCIENCE NEWS MARCH 25, 1978

EVIDENCE INDICATES BLACKHOLE
IN M-87 GALAXY
ASTRONOMY JULY 1978

SUPERMASSIVE OBJECT IN GALAXY M-87
SCIENCE NEWS   MAY 13, 1978

QUASARS
SCIENCE NEWS FEBRUARY 14, 1977

THE EVOLUTION OF QUASARS
MAARTEN SCHMIDT & FRANCES BELLO
SCIENTIFIC AMERICAN
COSMOLOGY MAY 1971

SOVIETS PROPOSE THEORY FOR
PHYSICAL BEHAVIOR OF QUASARS
ASTRONOMY   JULY 1977

FINDING OF MORE QUASARS DOUBTED
WILLIAM HINES
CHICAGO SUN-TIMES JANUARY 8, 1972

THE UNIVERSE

THE UNIVERSE
INTRODUCTION BY SIR BERNARD LOVELL
COLORPEDIA
THE RANDOM HOUSE ENCYCLOPEDIA
RANDOM HOUSE 1977

COSMOLOGY, HISTORY AND THEOLOGY
COSMOLOGY: MYTH OR SCIENCE ?
HANNES ALFVEN
YOURGRAU, BECK
PLENUM PRESS 1977

THE FIRST THREE MINUTES
STEVEN WEINBERG
BASIC BOOKS 1977

UNTIL THE SUN DIES
ROBERT JASTROW
W. W. NORTON 1977

WORLDS / ANTIWORLDS
HANNES ALFVEN

TEN FACES OF THE UNIVERSE
FRED HOYLE
W. H. FREEMAN   1977

THE COSMIC FRONTIERS OF
GENERAL RELATIVITY
WILLIAM J. KAUFMANN
LITTLE BROWN   1977

RELATIVITY AND COSMOLOGY
WILLIAM J. KAUFMANN
HARPER & ROW 1977

EXPLORING THE COSMOS
WITH ASTROPHYSICIST DAVID SCHRAMM
JACK HAFFERKAMP
THE READER / CHICAGO FEBRUARY 23, 1979

COSMIC MUSIC
NEW YORK TIMES   MARCH 22, 1978

HAVE ASTRONOMERS FOUND GOD ?
ROBERT JASTROW
NEW YORK TIMES   JUNE 25, 1978

IN THE BEGINNING WAS THE BANG ...
A BIG ONE
ROBERT JASTROW
LOS ANGELES TIMES JUNE 25, 1978

GAS BETWEEN GALAXIES ..
EXPANDING OR COLLAPSING ?
NEW YORK TIMES APRIL 27, 1977

WHIMPERING END IS SEEN FOR ALL THIS
WILLIAM HINES
CHICAGO TRIBUNE   NOVEMBER 15, 1978

MILKY WAY HURTLING THROUGH SPACE
NEW YORK TIMES NOVEMBER 14, 1977

FAR OUT: A SUPERGIANT CLUSTER OF GAL
CHICAGO SUN-TIMES   DECEMBER 14, 1979

LIGHT

PROBING THE UNIVERSE
NEWSWEEK   MARCH 12, 1979

THE YEAR OF DR. EINSTEIN
TIME   FEBRUARY 19, 1979

THE FIRST BILLION BILLION BILLION BILLIONTH
OF A SECOND ·· AND THEN
APRIL LAWTON
SCIENCE DIGEST MAY 1981

GENESIS REVEALED
ROBERT JASTROW
SCIENCE DIGEST WINTER 1979

TIME ZERO
MARK R. CARTRAM III
OMNI   NOVEMBER 1978

THE DARK SKY PARADOX AND
THE ORIGIN OF THE UNIVERSE
ASTRONOMY   SEPTEMBER 1978

UNIVERSE MIGHT BE 20 BILLION YEARS OLD
ASTRONOMY   JULY 1977

THE END OF TIME
BARRY PARKER
ASTRONOMY MAY 1977

SMOOTHING OUT THE UNIVERSE
DIETRICK THOMSEN
SCIENCE NEWS JANUARY 15, 1977

THE CURVATURE OF SPACE
A FINITE UNIVERSE
J. CALLAHAN
SCIENTIFIC AMERICAN
COSMOLOGY   AUGUST 1976

COLLAPSING UNIVERSE
ISAAC ASIMOV
SCIENCE DIGEST JUNE 1977

DOES THE UNIVERSE OSCILLATE?
JOHN GRIBBIN
ASTRONOMY AUGUST 1977

WILL THE UNIVERSE EXPAND FOREVER?
GOTT, GUNN, SCHRAMM & TINSLEY
SCIENTIFIC AMERICAN
COSMOLOGY   MARCH 1976

THE COSMIC BACKGROUND RADIATION
ADRIAN WEBSTER
SCIENTIFIC AMERICAN
COSMOLOGY   AUGUST 1974

COSMIC BACKGROUND RADIATION
INDICATES MILKY WAY TRAVELING
ONE MILLION M.P.H.
ASTRONOMY APRIL 1978

THE QUANTUM MECHANICS OF BLACK HOLES
S. HAWKING
SCIENTIFIC AMERICAN
COSMOLOGY   JANUARY 1977

MINI BLACK HOLES
BARRY PARKER
ASTRONOMY FEBRUARY 1977

AMAZING NUMERICAL COINCIDENCES
JOHN GLIEDMAN
SCIENCE DIGEST MAY 1981

LARGE NUMBERS, COSMIC COINCIDENCES,
AND GRAVITY
WAYNE HOPKINS
ASTRONOMY OCTOBER 1978

COSMOLOGY AGAINST THE GRAIN
DIETRICK E. THOMSEN
SCIENCE NEWS AUGUST 26, 1978

RECENT DEVELOPMENTS IN COSMOLOGY
FRED HOYLE
NATURE OCTOBER 9, 1965

COSMOLOGY BEFORE AND AFTER QUASARS
DENNIS SCIAMA
SCIENTIFIC AMERICAN
COSMOLOGY   SEPTEMBER, 1967

RIVAL COSMOLOGIES
BARRY PARKER
ASTRONOMY MARCH 1978

THE REDSHIFT PROBLEM
BARRY PARKER
ASTRONOMY SEPTEMBER 1978

BEYOND THE BLACK HORIZON
SCIENCE AND THE CITIZEN
SCIENTIFIC AMERICAN MAY 1982

THE MOST FEARED ASTRONOMER
ON EARTH ·· HALTON C. ARP
WILLIAM KAUFMANN III
SCIENCE DIGEST JULY 1981

MAYBE THEY ARE FASTER·THAN·LIGHT
SCIENCE NEWS   VOL 112   1977

REFLECTING ON SUPERLUMINAL LIGHT
SCIENCE NEWS   VOL 112   1977

KALEIDOSCOPE / SPACE·TIME
EDITED BY TONY JONES
NEWSWEEK / FOCUS...
MYSTERIES OF THE COSMOS JUNE 1980

THE WIZARD OF SPACE AND TIME
DENNIS OVERBYE
OMNI   FEBRUARY 1979

WE HAVE TO STAY IN OUR OWN UNIVERSE
SCIENCE NEWS MARCH 18, 1978

PHYSICS AT THE EDGE OF THE UNIVERSE
SCIENCE NEWS   DECEMBER 18, 1976

TAKING THE QUANTUM LEAP
FRED ALAN WOLF
HARPER & ROW 1981

SUPERGRAVITY AND THE UNIFICATION
OF THE LAWS OF PHYSICS
FREEDMAN & VAN NIEUWENHUITEN
SCIENTIFIC AMERICAN FEBRUARY 1978

A KEY NEW EXPERIMENT IN THE QUEST
TO UNIFY
SCIENCE DIGEST WINTER 1979

THE QUANTUM THEORY AND REALITY
BERNARD D'ESPAGNAT
SCIENTIFIC AMERICAN NOVEMBER 1979

VAN NOSTRAND'S SCIENTIFIC ENCYCLOPEDIA
VAN NOSTRAND

AN INTRODUCTION TO THE MEANING
AND STRUCTURE OF PHYSICS
LEON N. COOPER
HARPER & ROW 1968

QUARKS WITH COLOR AND FLAVOR
SHELDON LEE GLASHOW
SCIENTIFIC AMERICAN OCTOBER 1975

FUNDAMENTAL PARTICLES WITH CHARM
ROY F. SCHWITTERS
SCIENTIFIC AMERICAN FEBRUARY 1977

THE UPSILON PARTICLE
LEON M. LEDERMAN
SCIENTIFIC AMERICAN OCTOBER 1978

STALKING THE ELUSIVE GLUON
NEWSWEEK   SEPTEMBER 10, 1979

QUARK GLUE
SCIENCE AND THE CITIZEN
SCIENTIFIC AMERICAN NOVEMBER 1979

QUARKONIUM
ELLIOT D. BLOOM & GARY J. FELDMAN
SCIENTIFIC AMERICAN   MAY 1982

ATOMIC FIRST
OMNI JANUARY 1979

AND NOW CHARMED MOLECULES
PHYSICAL SCIENCES
SCIENCE NEWS MARCH 5, 1977

A DANCE OF PURE ENERGY
GARY ZUKAV
SCIENCE DIGEST WINTER 1979

THE ELEGANT SYMMETRY OF CRYSTALS
RODNEY C. EWING
NATURAL HISTORY FEBRUARY 1978

FIRST COLOR MOVIES CAPTURE ATOMS
IN MOTION
RONALD KOTULEK
CHICAGO TRIBUNE OCTOBER, 1978

THIS BOOK
WAS INSPIRED BY IDEAS
THAT CAME FROM
A LOT OF POPULARLY WRITTEN
BOOKS AND ARTICLES
ABOUT THE UNIVERSE
AND ITS
BEGINNINGS

THE PUBLISHER.. E·P·DUTTON
THE EDITOR ·· BILL WHITEHEAD
THE ART DIRECTOR·· NANCY ETHEREDGE
ASSISTANT EDITOR··KAREN STOUT
RESEARCH ASSISTANT. ·LINDA ZELENCIK

CARTOONS AND STORY BY BOB TOBEN